Wireless Power Transfer for Medical Microsystems

Tianjia Sun · Xiang Xie
Zhihua Wang

Wireless Power Transfer for Medical Microsystems

Tianjia Sun
Xiang Xie
Zhihua Wang
Institute of Microelectronics
Tsinghua University
Beijing
People's Republic of China

ISBN 978-1-4614-7701-3 ISBN 978-1-4614-7702-0 (eBook)
DOI 10.1007/978-1-4614-7702-0
Springer New York Heidelberg Dordrecht London

Library of Congress Control Number: 2013939335

Printed on acid-free paper

Springer is part of Springer Science+Business Media (www.springer.com)

Preface

In recent years, significant efforts have been dedicated to developing implantable electronic medical devices for biomedical applications. An incomplete list of such devices includes endoscopic capsule, artificial retinal prosthesis, implantable ECG recorder, artificial heart, and electrical stimulators. Traditionally, these devices are powered by implantable batteries or percutaneous cables. However, the limited energy budget of implantable batteries severely limits the system performance in terms of operating time, resolution, noise and so on, while the percutaneous cables make patients susceptible to infections.

To address the power deliver problem, an emerging trend is wireless power transfer. Such a solution can be dated back to a high-tension induction coil invented by Nikola Tesla. Since then, the wireless power transfer has been adopted in many electrical appliances including radio frequency identification and waterproofing products like electric shavers. Nowadays, researchers are taking efforts to employ it in biomedical applications. It delivers electrical energy to an implanted electrical device inside human body from a power source without man-made conductors.

The virtually unlimited power made possible by the wireless power transfer promises significant breakthroughs for implantable electronics. For example, surgical procedures will not be needed anymore to change implantable batteries in pacemakers. As a result, recent years witness a dramatic growth of the wireless power transfer technology. However, there are still so many challenges to design a wirelessly powered implant, like power transfer efficiency, power stability, and sizes of power antennas and circuits. This book develops a systematic treatment to this subject. We introduce in-depth current antenna and circuit solutions to meet these challenges. Two wirelessly powered capsule endoscopic systems are given as design cases. We believe wirelessly powered implants will change the landscape of future healthcare devices. We hope this book helps to develop a foundation for this promising technology.

Tianjia Sun
Xiang Xie
Zhihua Wang

Acknowledgments

There are many people that helped make this book a reality that deserve our thanks and recognition. First, we would like to thank many professors that we had the pleasures of working with. They include Prof. Guolin Li and Yangdong Deng for reviewing this research work and providing valuable feedback. We are also very grateful to Prof. Chun Zhang, Hong Chen, Hanjun Jiang, Baoyong Chi, and Woogeun Rhee for their leadership, support, and encouragement. Moreover, we would like to thank other members in our group. The first is Yingke Gu. He designed and tested the wireless capsule endoscopy system with us together. Additionally, we would like to show our gratitude to Xiaomeng Li, Yadong Huang, Huanhuan Li, and Xunxun Zhu for their hard working on many valuable circuit, antenna, and packaging technologies. We will always remember with fondness of the years that we worked side by side. We would also like to recognize the help from Dr. Liyuan Liu, now at the Institute of Semiconductors of Chinese Academy of Science, and Dr. Bo Zhou, now at Beijing Institute of Technology. They helped us in starting our designs at the very beginning. Lastly, we are grateful to Springer for publishing this book.

Contents

Chapter 1
Introduction

Abstract In recent years, the improving in the function, performances, and operating time of kinds of biomedical microsystems (e.g. nerve stimulators, implantable monitors, endoscopic capsules, etc.) is pushing up their power requirement. Traditional implantable batteries and percutaneous cords are suffering from low reliability and high infection risks. Accordingly, it leads to interests in an old electromagnetic technology, the wireless power transfer (WPT), which was invented over 100 years ago. The WPT is promising way to safely provide more energy or enable longer lifetime for modern biomedical applications. Believing the WPT is a perfect way out, researchers are devoting significant efforts to develop the technology, especially in recent five years. This book presents in-depth the design of the wireless power transfers applied in biomedical microsystems. In the first chapter, the motivation of the book, the brief history of the WPT, the catalogue of the WPT, and the target applications are to be introduced.

1.1 Motivations

In recent years, significant efforts have been dedicated to developing biomedical microsystems. These microsystems include endoscopic capsules [1–6], artificial retinal prosthesis [7], pacemakers [8, 9], artificial hearts [10], electrical stimulators [11], lab-on-a-chip [12], and so on.

The traditional approach to supplying power to these devices is to use implantable batteries or percutaneous cables. The limited energy budget of implantable batteries severely limits the system performance in terms of the operating time, the resolution, and so on, while the percutaneous cable makes patients susceptible to infections and also results in unreliability problems. For example, the battery powered capsule endoscopy can operate approximately 6–8 h [4]. However, for some patients, the capsule might stay in the digestive track for more than 8 h. Another example is the pacemaker. Typically, a battery in a

T. Sun et al., *Wireless Power Transfer for Medical Microsystems*,

DOI: 10.1007/978-1-4614-7702-0_1,

pacemaker can last around 5–10 years [13]. Doctors need to monitor the pacemaker. Surgical procedure will be required to change the battery for the next 5–10 years.

In recent years, there is another emerging power option for the biomedical microsystems. It's the wireless power transfer (WPT), which is right the topic of this book. Conventionally, the wireless power transfer is used in electrical systems including radio frequency identification (RFID) [14] and waterproofing products like the electric shaver. Nowadays, the technology of the WPT is adopted to transfer power from the electrical equipment in vitro to the implantable microsystem in vivo.

The use of the wireless power transfer is a significant breakthrough for modern implants. It offers an unlimited remote power source and it improves system overall features. Because the power can be wirelessly transferred into human body, the usage and the characteristic of medical systems can be changed. For instance, as mentioned, the battery in the conventional pacemaker has to be changed in a period of 5–10 years [13]. Thanks to the wireless power transfer, the pacemaker can be wirelessly recharged in the future, so surgeries can be avoided for patients.

On a global scale, the wireless power transfer is developing at an amazing speed. Figure 1.1 shows the numbers of published researches for the usage in biomedical microsystems in recent ten years. The technology of the wireless power transfer is already an extremely important development direction.

In these researches, many kinds of specialized antenna and circuit techniques have been developed to improve the transfer and conversion efficiencies, reduce system size, promote system stability, and so on. In the future, the performance-enhanced implants may aid human recovery from trauma, regain sight, and reduce pain. We believe it will play an important role in the future. Accordingly, the main topic of this book is to present in-depth the recent technologies to design wireless power transfer for medical applications. This book introduces:

- Brief WPT history, categories and modern applications
- WPT fundamentals, system design components, challenges and considerations including distance, efficiency, size, electromagnetic safety, and so on
- The state-of-the-art technologies of power antennas, converters, and management adopted in the WPT
- Detailed design cases of wirelessly powered medical microsystems

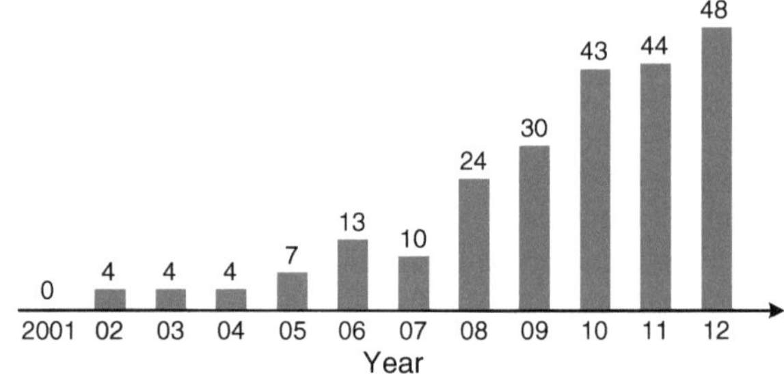

Fig. 1.1 The rapid development of the WPT applied in biomedical applications, the number is indexed by using key words "Wireless Power Transfer and Biomedical" in IEEE Xplore

1.2 Brief History

The wireless power transfer originated from almost 200 years ago. The timeline below describes the development of the wireless power transfer since the very beginning.

1826–1831 Andre-Marie Ampere developed Ampere's circuital law [15], which shows the electric current produces a magnetic field. Michael Faraday developed Faraday's law of induction [15], which describes the electromagnetic force can be induced in a conductor by a time-varying magnetic flux.

1891–1894 Nikola Tesla demonstrated the first wireless power transfer by means of electrostatic induction using a high-tension induction coil before the American Institute of Electrical Engineers at Columbia College [16]. Tesla demonstrated the wireless illumination of phosphorescent lamps of his design at the World's Columbian Exposition in Chicago [17]. He also demonstrated wireless transmission of signals before a meeting of the National Electric Light Association in St. Louis [18, 19] (Fig. 1.2).

1894–1899 Tesla lighted incandescent lamps wirelessly at the 35 south Fifth Avenue laboratory in New York City by resonate inductive coupling [20]. Hutin and LeBlanc received U.S. Patent 527,857 describing a system for power transmission at 3 kHz [21]. Jagdish Chandra Bose ringed a bell at a distance using electromagnetic waves and also ignited gunpowder [22]. Marconi demonstrated radio transmission over a distance of 1.5 miles [23]. Tesla demonstrated wireless transmission over a distance of about 48 km in 1896 [24] and continued his wireless power transfer research.

1904–1917 At the St. Louis World's Fair, a prize was offered for a successful attempt to drive a 0.1 horse power airship motor by energy transmitted through space at a distance of at least 30 m [25].

Fig. 1.2 Wireless transmission of power demonstrated by Tesla during his 1891 lecture on high frequency and potential [16], [49]

Fig. 1.3 From the Wardenclyffe plant, Tesla hoped to demonstrate wireless transmission of electrical energy across the Atlantic [50]

Meanwhile, Tesla's Wardenclyffe Tower was built in 1901 and it was a commercial and a scientific demonstration of trans-Atlantic wireless telephony, broadcasting, and wireless power transmission. It was never fully operational and the tower was demolished during the World War I because US government feared German spies were using it (Fig. 1.3).

1926–1968 Hidetsugu Yagi published their first paper on the Yagi antenna [26]. William Brown published an article exploring possibility of microwave power transfer [27]. A system used to wirelessly transmit solar energy captured in space was proposed [28], which is recognized as the first solar power satellite.

1973–1998 The word's first passive RFID system was demonstrated at Los-Alamos National Lab [29] in 1973. Goldstone Deep Space Communications Complex did experiments in tents of kilowatts [30]. RFID tags can be powered by electrodynamic induction over a few feet.

2007–2013 Using strongly magnetic resonance, the WiTricity research group led by MIT wirelessly powered a 60 W bulb with 40 % power efficiency at a distance of 2 m [31]. Intel reproduced the original 1894 implementation of electrodynamic induction by wirelessly powering a nearby light bulb with 75 % efficiency [32]. Sony showed a wireless electrodynamic-induction powered TV set, 60 V over 50 cm [33]. Haier showed a wireless LCD TV at CES 2010 using researched Wireless Home Digital Interface [34]. In this period, our research team focused on the wireless power transfers for biomedical applications [2, 3, 35–38].

1.3 Category for the Wireless Power Transfer

After many years of development, there have been many types of wireless power transfer and they can be cataloged by many ways, for example by the efficiency, power level, size, operating frequency, transmission distance, and so on. Here, in this book, we classify the category of the wireless power transfer by the essential physical working principle. Figure 1.4 shows the category.

As above figure shows, there are two basic sorts. They are the near-field transfers and the far-field transfers. The difference between the two transfer ways is the characteristics of electromagnetic fields changing with distance from the charges and currents producing the electromagnetic field. When the resonate frequency of the changing electromagnetic field is relative low (like 1 MHz) and the transfer distance is relative short (like 10 cm), it belongs to the near-field transfer and vice versa. The boundary between the two kinds of transfers is vaguely defined. For transmitters and receivers in diameters shorter than half of the operating wavelength, the near field is the region within a radius of wavelength ($r<\lambda$), while the far-field is the region out of a radius of two wavelengths ($r > 2\lambda$). The middle region between is called a transition zone. For transmitters and receivers in diameter larger than a half-wavelength, the near and far field transfers are defined by the Fraunhofer distance [39]:

$$d = \frac{2D^2}{\lambda} \tag{1.1}$$

where D is the dimension of the largest antenna of the power transmitter and the receivers, λ is the wavelength of the electromagnetic wave. Typically, the near-field transfer has higher power transfer efficiency over the far-field transfer. Using longer wave length, the near-field transfer is easier to generate diffraction when the electromagnetic wave encounters human body. Accordingly, we pay much more attention on the near-field transfer.

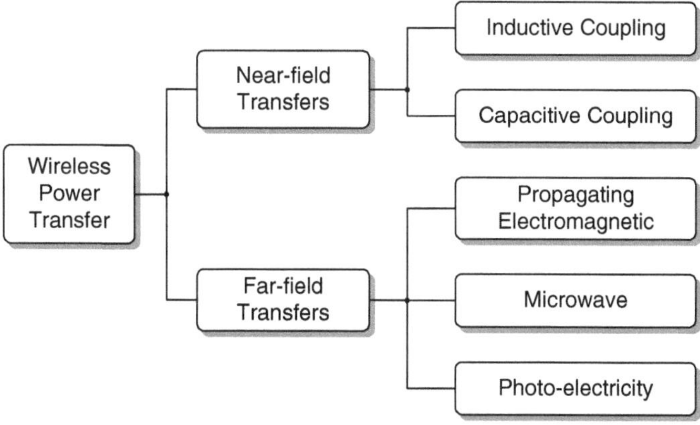

Fig. 1.4 The category of the wireless power transfer

	Freq	Directivity	Range	Penetrability	Efficiency
Inductive Coupling	Low Hz~MHz	Weak	Short	Strong	High
Capacitive Coupling	Low Hz~MHz	Weak	Short	Strong	High
Propagating Electromagnetic	Medium MHz~GHz	Medium	Medium	Medium	Medium
Microwave	High GHz~THz	Strong	Long	Weak	Low
Photo-electricity	High >THz	Strong	Long	Weak	Low

Fig. 1.5 A comparison among the wireless power transfers

Both the near-field and far-field transfers have subtypes. The near-field transfers include the inductive coupling and the capacitive coupling. The far field transfers include the propagating electromagnetic transfer, the microwave transfer, and the photo-electric transfer. The characteristics of the five kinds of transfers are summarized in the Fig. 1.5. According to the demand for biomedical applications, there is little requirement on the directivity and transfer range. However, the penetrability and power efficiency are extremely important. As a consequence, the inductive coupling and the capacitive coupling seem to be more suitable for biomedical applications because their higher efficiency and stronger penetrability.

Figure 1.6 shows the detail of the inductive and capacitive coupling. The capacitive coupling transfers energy through alternating electric field, while the inductive delivers energy using alternating magnetic field. Compared to the

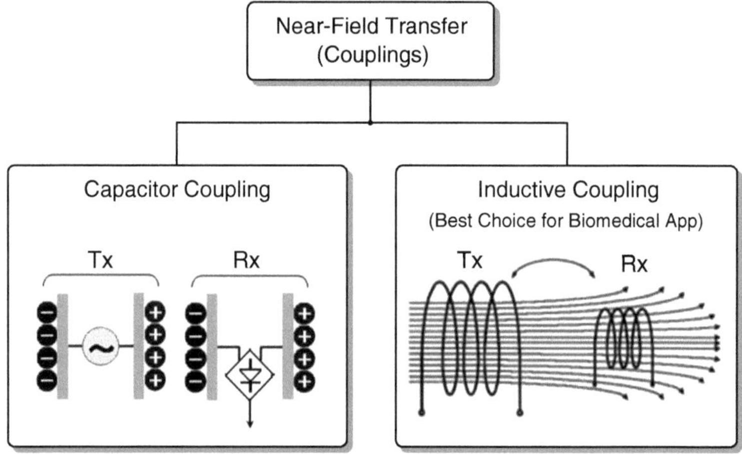

Fig. 1.6 A comparison between the capacitor and inductive couplings

electric field, magnetic field causes much less adverse effects on human body, which makes the inductive coupling the best choice for biomedical applications.

1.4 Target Applications

The wireless power transfer has been applied in many modern applications. The most well-known application is the Radio Frequency Identification (RFID) [14]. The RFID is a wireless non-contact system that uses radio frequency electromagnetic fields to transfer data from a tag attached to an object for the purpose of automatic identifications and tracking. The RFID systems can be classified into passive systems and active systems. In a typical passive system, there are one reader and at least one passive tag as shown in Fig. 1.7. The reader uses local power source and emits radio waves. The tag contains electronically stored information. When the tag comes near to the reader, it picks up energy from the radio waves and sends its information to the reader. Systems like this can be very useful in daily life, for example the access controlling system and the automatic cargo tracking in logistics management.

The wireless power transfer can be adopted in many waterproofing products like a wireless electric toothbrush or a wireless electric shaver. As Fig. 1.8 shows,

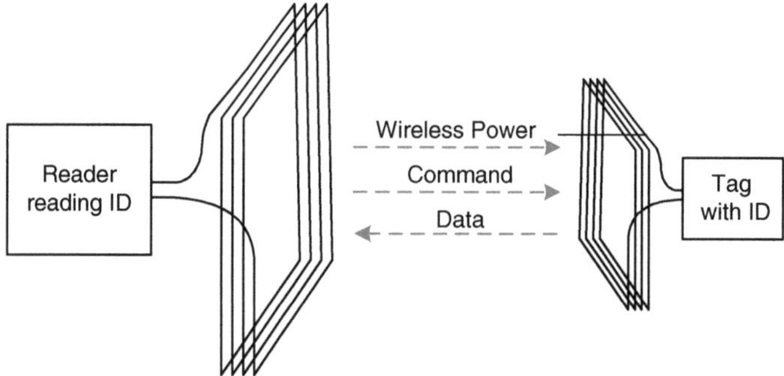

Fig. 1.7 A RFID system based the wireless power transfer

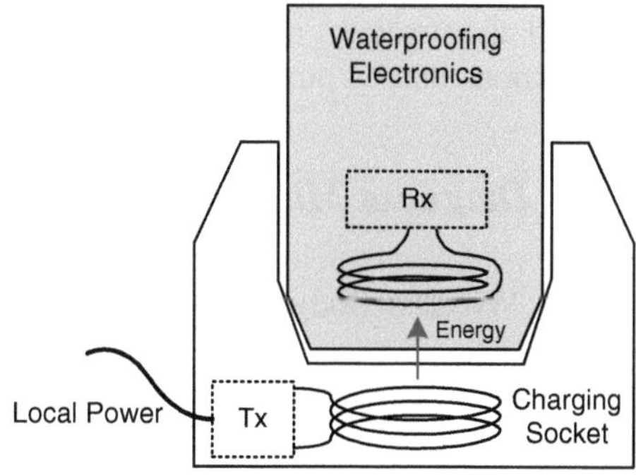

Fig. 1.8 A waterproofing electric shaver based on wireless power transfer

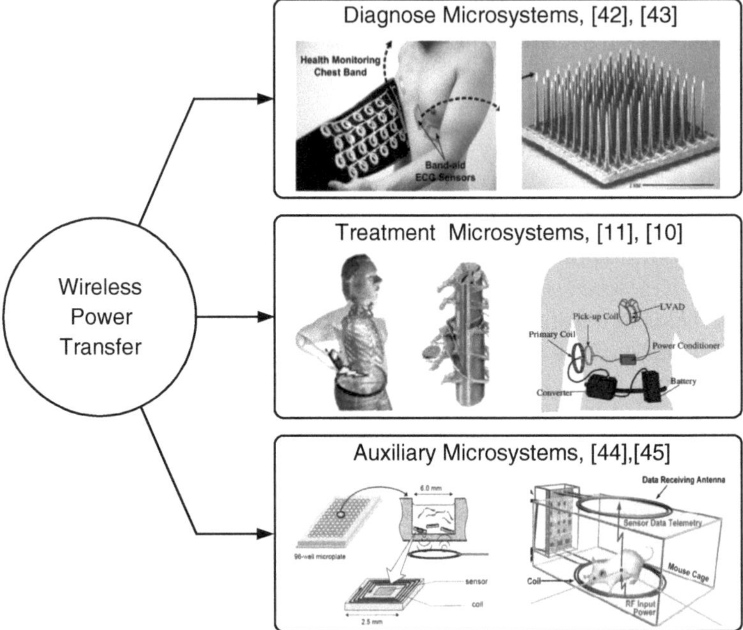

Fig. 1.9 Three types of target applications

to make the product waterproofing, there is no metal contact on the surface of the product. The power is wirelessly induced from a socket to the internal circuits.

Besides the above applications, as the topic of this book, the wireless power transfer has been also adopted in many biomedical applications. Figure 1.9 shows the catalog and six examples.

By the catalog in Fig. 1.9, the wireless power transfer is adopted to three kinds of target biomedical microsystems including diagnostic microsystems, treatment microsystems, and auxiliary microsystems. In the figure, two examples are given for each catalog. A body sensor system [40] and an implantable neural recording array [41] are presented for diagnostic microsystems. A dorsal stimulator [11] and an artificial heart [10] are given as typical treatment microsystems. A wireless lab-on-a-chip [42] and a wireless implantable system for mice experiment [43] are showed as auxiliary medical microsystems. We introduce the three types of applications one by one as follows.

1.4.1 Diagnose Microsystems

1.4.1.1 Wireless Body Sensor Nodes

The rapid growth in wearable sensors, low power integrated circuits and wireless communications have enabled a new generation of biomedical electronics, which is the wireless body sensors. Typically, these sensors can constitute a wireless network around human body to monitor and deliver the healthcare information about the body. Regular body sensors use button batteries as the power source. On one hand, the size of integrated circuits is usually much smaller than a battery, thus the size of the battery determines the system size. On the other hand, the energy budget of the battery is very limited, so the lifetime of sensor system is usually unsatisfactory. To address this problem, many previous researches have turn to use the wireless power transfer as the power source. For some systems, the power is emitted from an active transmitter. These systems usually require relative large power, like milliamperes. For other systems, there are no specially designed power transmitters. The sensors just harvest the electromagnetic energy generated by other applications in free space. For example, some sensors collect the high-frequency alternating electromagnetic energy generated by mobile phones base stations. Typically, these sensors have quite small power consumption.

1.4.1.2 Wireless Nerve Sensing System

The wireless power transfer can be applied to another type of diagnostic medical microsystem, which is the nerve sensing or neural recording system. Typically, a nerve signal sensor will be placed in brain. The sensor monitors a group of nerve signals. These signals can be recorded, processed, or transmitted out of brain to a computer. Figure 1.10 shows a physically disabled person is using this technology to control a mechanical arm to drink a cup of water. First, the wireless power is transmitted from an external device to the nerve sensor in the brain. Second, the nerve signals are wirelessly transmitted out and processed. And finally, the mechanical arm is controlled to raise the cup.

The system above is also called brain-computer-interface (BCI) [41, 44, 45]. In the future, it would help people to assist, augment, or repair more human cognitive or sensory-motor functions.

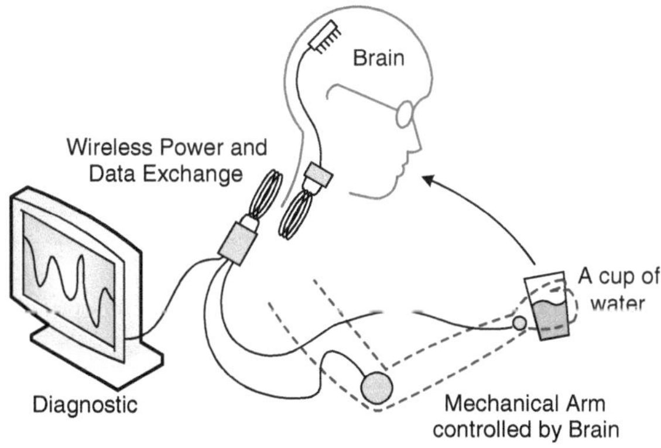

Fig. 1.10 Wireless power transfer used in a wireless neural sensing system

Another good example might be the cochlear implant [46]. The cochlear implant is a surgically implanted electric device that provides a sense of sound to a person who is profoundly deaf. Typically, a cochlear implant is composed of an implantable part with wireless power receiver and an external part with power transmitter.

1.4.2 Treatment Microsystems

1.4.2.1 Batteryless Nerve Stimulation

Electrical stimulation to nerve tissue can be very useful in treatments for spinal cord injury, stroke, sensory deficits, and neurological disorders. The electrical stimulation can be operated by implantable stimulators. Conventional stimulators use batteries as the power source, which seriously limits the lifetime of the stimulators. Researchers have proposed batteryless nerve stimulators [11] as Fig. 1.11 shows.

For example, a sacral nerve stimulator can be implanted in the buttocks of people who have problems with bladder or bowel control. The stimulator can be powered by wireless energy delivered from an external device, so the stimulator does not suffer from the limited energy budget of battery. When external device works, the stimulator gets energy and electrically stimulates the sacral nerve to control the bladder or bowel. Accordingly, patients are able to control their bladder behavior on-demand using the system.

1.4.2.2 Wirelessly Powered Artificial Organs

An artificial organ is a man-made device that is implanted or integrated into a human to replace a natural organ for purpose of restoring a specific function so the

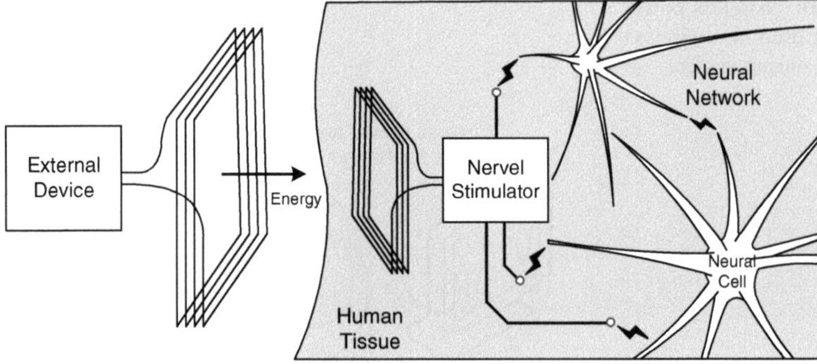

Fig. 1.11 Wireless power transfer used in a nerve stimulator

patients may return a normal life. Some artificial organs use electrical power, for example the artificial heart and the pacemaker.

The traditional way to provide power to these devices is implantable battery or percutaneous cord. However, the battery has a limited energy budget and percutaneous has reliability problems and the infection risks. The wireless power transfer is right a perfect solution to these problems. Previous research [10] showed a wireless power transfer system for left ventricular assist device (LVAD) and artificial heart. A primary-side wireless power transmitter in vitro is used to produce a high frequency alternating magnetic field. A secondary-side wireless power receiver in vivo picks up and offers the energy to artificial organs.

1.4.3 Auxiliary Microsystems

1.4.3.1 Wireless Power Transfer for Lab-on-a-Chip

In recent years, the concept of the lab-on-a-chip [12, 47, 48] draws much attention. A lab-on-a-chip is a device that integrates laboratory functions on an integrated chip. Common functions include DNA sequencing, biochemical analysis, synthesis, and so on. These devices may be implanted into human body or soaked in a test tube. The size of the lab-on-a-chip is typical several square millimeters. Consequently, the wireless power transfer is quite a perfect power solution for such small microsystems. Figure 1.12 shows a typical a lab-on-a-chip system using the wireless power transfer.

The wireless power is generated by an external device. The lab-on-a-chip uses on-chip spiral coils to pick up the energy and convert the picked AC energy to DC energy for designed functions. There are two special problems for the WPT in the lab-on-a-chip. First, the quality factor of the on-chip coil is usually much lower than the quality factor of off-chip inductors. Second, the size of the on-chip coil is usually under millimeters. Consequently, to transfer energy efficiently is quite a challenge. Other chapters of this book will introduce the state-of-the-art methods to address the problems.

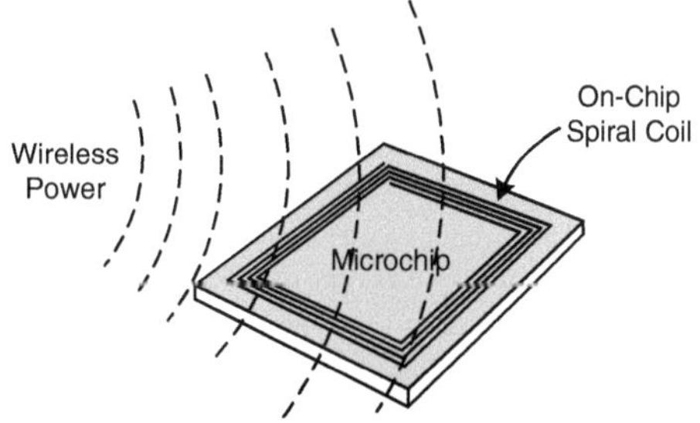

Fig. 1.12 Wireless power in Lab-on-a-chip

1.4.3.2 Wireless Animal Test Systems

For most pharmaceutical factories and doctors, animal tests are necessary before clinical trials on human body. The animal tails provide safety and efficacy data about the adverse drug reactions and adverse effects. Depending on the type of researches or products, deferent types of physiological parameters should be monitored during the animal tests. For some type of parameters, continuous monitoring is preferred. Previous research [43] proposed a continuous and implantable blood pressure monitoring system for the mice test. In the system, power is wirelessly transmitted outside of mice cage. A batteryless blood pressure monitor in mice receives the energy and transmits the pressure data wirelessly out. The advantage of the system is the mice will not be disturbed after the experiment is started.

1.5 Organization of the Book

For ease of comprehension, this book has a very straightforward structure with total 7 chapters. Figure 1.13 illustrates the overall structure.

Chapter 1 motivates the book, introduces the brief history and the catalog of the wireless power transfer, presents target applications, and outlines the rest of the book.

Chapter 2 presents the system level design. The systematic design methods, system components, design challenges, and safety considerations are illustrated.

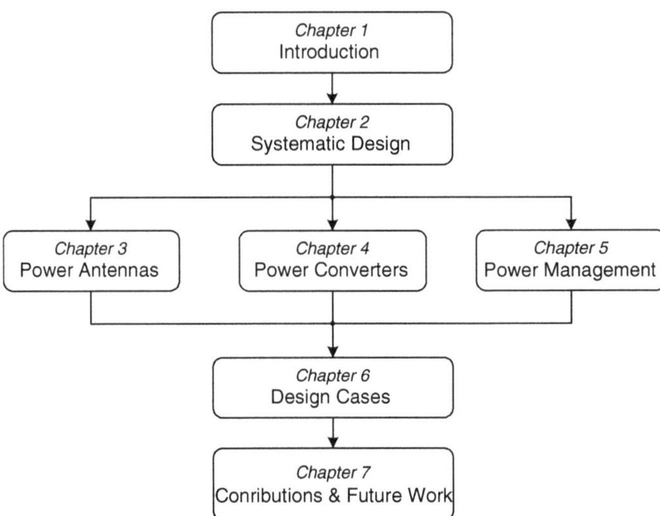

Fig. 1.13 Organization of the book

Chapter 3 covers the first type of system components, the power antenna. The State-of-the-art power antenna techniques are investigated and compared.

Chapter 4 covers the second type of system components, the power converter. Featured power conversion circuits are given for different types of applications.

Chapter 5 covers the last type of system components, the power management unit. Several power management techniques are proposed.

Chapter 6 demonstrates two detail design cases. They use the techniques introduced in previous chapters.

Chapter 7 summarizes the book's main contributions including design methods and design techniques and takes a look into future.

References

1. Lenaerts, B., & Puers, R. (2007). An inductive power link for a wireless endoscopy. *Biosensors and Bioelectronics, 22*(7), 1390–1395.
2. Sun, T., Xie, X., Li, G., et al. (2012). A two-hop wireless power transfer system with an efficiency-enhanced power receiver for motion-free capsule endoscopy inspection. *IEEE Transactions on Biomedical Engineering, 59*(11), 3247–3254.
3. Sun, T., Xie, X. & Li, G., et al. (2011). *An omnidirectional wireless power receiving IC with 93.6% efficiency CMOS rectifier and Skipping Booster for implantable bio-microsystems, A-SSCC* (pp. 185–188).
4. Xie, X., Li, G., Chen, X., Li, X., et al. (2006). A low-power digital IC design inside the wireless endoscopic capsule. *IEEE Journal of Solid-State Circuits, 41*(11), 2390–2400.
5. Chen, X., Zhang, X., Zhang, L., et al. (2009). A wireless capsule endoscope system with low-power controlling and processing ASIC. *IEEE Transactions on Biomedical Circuits and Systems, 3*(1), 11–22.
6. Carpi, F., Kastelein, N., Talcott, M., et al. (2011). Magnetically controllable gastrointestinal steering of video capsules. *IEEE Transactions on Biomedical Engineering, 58*(2), 231–234.
7. Naganuma, H., Kiyoyama, K., & Tanaka T. (2012). *A 37 × 37 pixels artificial retina chip with edge enhancement function for 3-D stacked fully implantable retinal prosthesis, BioCAS*, (pp. 212–215).
8. Wong, L. S., Hossain, S., Ta, A., et al. (2004). A very low-power CMOS mixed-signal IC for implantable pacemaker applications. *IEEE Journal of Solid-State Circuits, 39*(12), 2446–2456.
9. Lee, S.-Y., Su, M. Y., Liang, M.-C., et al. (2011). A programmable implantable microstimulator SoC with wireless telemetry: application in closed-loop endocardial stimulation for cardiac pacemaker. *IEEE Transactions on Biomedical Circuits and Systems, 5*(6), 511–522.
10. Si, P., Hu, A. P., Malpas, S., et al. (2008). A frequency control method for regulating wireless power to implantable devices. *IEEE Transactions on Biomedical Circuits and Systems, 2*(1), 22–29.
11. Chiu, H.-W., Lin, M.-L., Lin, C.-W., et al. (2010). Pain control on demand based on pulsed radio-frequency stimulation of the Dorsal root Ganglion using a batteryless implantable CMOS SoC. *IEEE Transactions on Biomedical Circuits and Systems, 4*(6), 350–359.
12. Schluter, M., Mammitzsch, S., & Lilienhof, H. J. (2005). *Detection of a cardiac infarct on a disposable lab on a chip device: 3rd IEEE/EMBS Special Topic Conference on Microtechnology in Medicine and Biology* (pp. 69–71).
13. Ghallab, Y. H., & Badawy, W. (2005). *A novel CMOS lab-on-a-chip for biomedical applications, ISCAS* (pp. 1346–1349).

14. Finkenzeller, K. (2010). *RFID handbook: fundamentals and applications in contactless smart cards, radio frequency identification and near-field communication*. New York: Wiley.
15. Thidé, B. (2004). *Electromagnetic field theory*. Sweden: Upsilon Books.
16. Tesla, N. (2007). *Experiments with alternate currents of very high frequency and their application to methods of artificial illumination*. New York: Wilder Publications.
17. Barrett, J. P. (1894). *Electricity at the Columbian exposition* (pp. 168–169). Chicago: R. R. Donnelley.
18. Tesla, N. (2006). *On light and other high frequency phenomena*. Rockville: Wildside Press.
19. Tesla, N. (2003). *Nikola Tesla 1856–1943*. Belgrade: Nolit.
20. Tesla, N. (1892). *Experiments with alternate currents of high potential and high frequency: IEEE London*.
21. "Transformer System for Electric Railways". United States Patent Office, Retrieved, 2010.
22. Emerson, D. T. (1997). *The work of Jagadish Chandra Bose: 100 years of mm-wave research: National radio astronomy observatory*.
23. Marconi Wireless Tel. Co. v. United States, 320 U.S.1, US Supreme Court Center, 1943.
24. Anderson, L. I., (2002). *N. Tesla, Nikola Tesla on his work with alternating currents and their application to wireless telegraphy, telephony, and transmission of power: An extended interview: 21st Century Books*.
25. Wikipedia, http://en.wikipedia.org/wiki/Wireless_power.
26. Yagi, H. (1928). Beam transmission of ultra-shortwaves. *Proceedings of the IRE, 16*, 715–740.
27. Brown, W. R. (1961). A survey of the elements of power transmission by microwave beam. *IRE, 9*(3), 93–105.
28. Glaser, P. E. (1968). Power from the Sun: Its future. *Science, 162*(3856), 857–861.
29. Landt, J. (2005). The history of RFID. *Potentials, IEEE, 24*(4), 8–11.
30. Shinohara, N. (2006). *Wireless power transmission for solar power satellite*. Space Solar Power Workshop, Georgia Institute of Technology.
31. Kurs, A., Karalis, A., Moffatt, R., et al. (2007). Wireless power transfer via strongly coupled magnetic resonances. *Science, 317*, 83–86.
32. Intel.(2008). Intel imagines wireless power for your laptop, http://www.tgdaily.com.
33. Sony. (2009). Sony develops highly efficient wireless power transfer system. http://presscentre.sony.eu.
34. Haier. (2010). Haier's wireless HDTV lacks wires. http://www.engadget.com.
35. Sun, T., Xie, X., Li, G. L., et al. (2012). Integrated omnidirectional wireless power receiving circuit for wireless endoscopy. *Electronics Letter, 48*(15), 907–908.
36. Sun, T. J., Xie, X., Li, G. L., et al. (2012). Rectigulator, a hybrid of rectifiers and regulators for miniature wirelessly powered bio-microsystems. *Electronics Letter, 48*, 1181–1182.
37. Sun, T., Xie, X., Li G. L. et al. (2010). *A wireless energy link for endoscopy with end-fire helix emitter and load-adaptive power converterm, APCCAS* (pp. 32–35).
38. Sun, T., Xie, X., Li G. L. et al. (2010). *An asymmetric resonant coupling wireless power transmission link for Micro-Ball Endoscopy, EMBC,* (pp. 6531–6534).
39. Balanis, C. A. (1982). *Antenna theory: analysis and design*. New York: Wiley.
40. Yoo, J., Yan, L., Lee, S., et al. (2010). A 5.2 mW self-configured wearable body sensor network controller and a 12uw 54.9% efficiency wirelessly powered sensor for continuous health monitoring system. *IEEE Journal of Solid-State Circuits, 45*(1), 178–188.
41. Harrison, R. R., Watkins, P. T., Kier, R. J., et al. (2007). A low-power integrated circuit for a wireless 100-electrode neural recording system. *IEEE Journal of Solid-State Circuits, 42*(1), 123–133.
42. Yazawa, Y., Oonishi, T., & Watanabe, K., et al. (2005). *A wireless biosensing chip for DNA detection, ISSCC* (pp. 562–617).
43. Cong, P., Chaimanonart N., & Ko, W. H., et al. (2009). *A wireless and batteryless 130 mg 300 μW 10b implantable blood-pressure-sensing microsystem for real-time genetically engineered mice monitoring, ISSCC* (pp. 428–429).

44. Serby, H., Yom-Tov, E., et al. (2005). An improved P300-based brain-computer interface. *IEEE Transactions on Neural Systems and Rehabilitation Engineering, 13*(1), 89–98.
45. Schalk, G., McFarland, D. J., Hinterberger, T., et al. (2004). BCI2000: a general-purpose brain-computer interface (BCI) system. *IEEE Transactions on Biomedical Engineering, 51*(6), 1034–1043.
46. Zeng, F.-G., Rebscher, S., Harrison, W., et al. (2008). Cochlear implants: system design, integration, and evaluation. *IEEE Reviews in Biomedical Engineering, 1*, 115–142.
47. Ahn, C. H., Choi, J.-W., Beaucage, G., et al. (2004). Disposable smart lab on a chip for point-of-care clinical diagnostics. *Proceedings of the IEEE, 92*(1), 154–173.
48. Cho, S. K., Moon, H., & Kim, C.-J. (2003). Creating, transporting, cutting, and merging liquid droplets by electrowetting-based actuation for digital microfluidic circuits. *Journal of Microelectro Mechanical Systems, 12*(1), 70–80.
49. Norrie, H. (2009). *Induction coils: how to make, use and repair them*: *Wireless Telegraphy and Practical Information*.
50. Tesla, N. (1908). The future of the wireless art. In: Wireless Telegraphy and Telephony. Walter W. Massie & Charles R. Underhill, pp. 67–71.

Chapter 2
Systematic Designs

Abstract A successful systematic design is extremely important for wireless power transfer. It helps to clarify the features that are the most important and that can be given up. It also helps to develop antennas and circuits techniques for the system. In this chapter, we focus on the system level design of the wireless power transfer for biomedical applications. Although biomedical applications may be variable, we try to model, extract, and discuss the common features for their power transfers. First, the basic working principle of the inductive coupling is briefly depicted. Second, a unified systematic model is proposed for the transfers. Three types of the system components, including the power antennas, the power converters, and the power management are to be described. Third, typical challenges in the systematic design are to be summed up. The most challenging part is to trade off among various characteristics. At last, the electromagnetic exposure to human body is discussed, which is a necessary consideration in the systematic design.

2.1 Fundamentals

Before designing systems, we introduce the fundamental working principles of the wireless power transfer. As introduced in the Chap. 1, there are many types of the wireless power transfers. According to the demand for biomedical applications, stronger penetrability, higher power efficiency, and safer performance are the most critical features. As a consequence, the inductive coupling is the best choice for the biomedical applications. Therefore, the fundamentals discussed in this section are all based on the inductive coupling. These fundamentals mainly include the basic concepts of the time-varying Electromagnetic field, like the Ampere's circuital law [1], the Biot-Savart law [1], and the Faraday's law of induction [1]. Understanding these fundamental working principles is important for further discussion later.

T. Sun et al., *Wireless Power Transfer for Medical Microsystems*,
DOI: 10.1007/978-1-4614-7702-0_2,

2.1.1 Basic Laws

Considering the inductive coupling is the most frequently adopted transfer type in the biomedical applications, we don't distinguish the "wireless power transfer" and the "inductive coupling" anymore in the rest of the book.

A typical inductive coupling system consists of a primary power transmitter and a secondary power receiver. The transmitter adopts power converters to convert DC energy to AC energy and delivers the AC energy to a power transmitting antenna or coil. The receiver employs power receiving antenna or coil to pick up AC energy in space and converters the AC energy back to the DC energy. In some applications, the operating frequency of the time-varying electrometric field may be as low as 50–60 Hz, or as higher as 2.4 GHz. However, in the most of biomedical applications, the operating frequency is in the range from 100 kHz to 50 MHz. Comparing to the typical transmission distance in range from 1 to 10 cm in biomedical applications, the corresponding wave length of the electromagnetic field is relatively much longer. Therefore, the transfer can be viewed as a near field progress. The whole progress can be analyzed by using basic electromagnetic laws. To clearly illustrate the transfer progress, we start from the primary side by assuming there is a time-varying current in the transmitting coil.

(a) Magnetic flux density Generated by the Primary Coil

Suppose there is a continual flow of charges which is constant in time and the charge neither accumulates nor depletes at any point, the Biot-Savart law [1] gives out the magnetic flux density generated by the flow of charges.

$$B = -\frac{\mu_0}{4\pi} \oint_l \frac{Idl \times e_r}{r^2} \tag{2.1}$$

where r is the full displacement vector from the wire element to the point, at which the field is being computed, e_r is the unit vector of r. Idl is linear-current-element in the wire, and μ_0 is the magnetic constant.

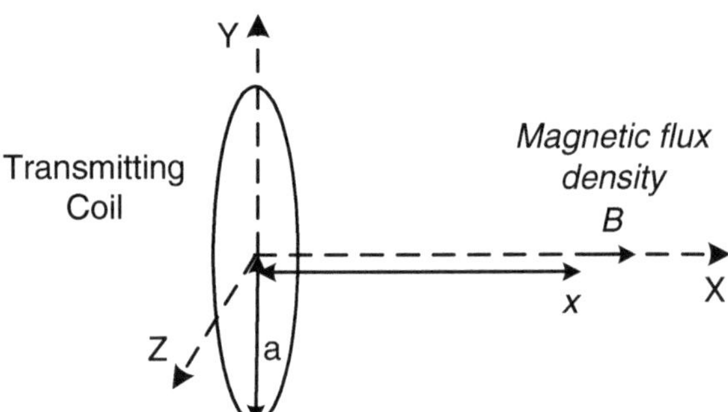

Fig. 2.1 Magnetic field generated by a circular transmitting coil

Typically in the wireless power transfer, the transmitting coil is a circular coil as shown in the Fig. 2.1.

For the circular coil, the generated magnetic flux density B at the point x in the Fig. 2.1 can be expressed by:

$$B_x = \frac{\mu_0 N I a^2}{2(a^2 + x^2)^{3/2}} e_x \tag{2.2}$$

where N is the coil's turns, I is the current in each turn, a is the radius of the circular coil, x is the distance from the center of the coil to the point x, and e_x is the unit vector of axle X. If the current I in the transmitting coil is not constant but time-varying, the generated magnetic flux density B_x would also change over time.

(b) Induced voltage in the Secondary Coil

Suppose there is another circular coil in space as shown in the Fig. 2.2.

The total time-varying magnetic flux Φ_{m} crossing the secondary coil can be expressed by:

$$\Phi_{\mathrm{m}} = \int_s B \cdot dS \tag{2.3}$$

where B is the magnetic flux density generated by the primary coil, and S is the surface of the secondary coil. According to the Faraday's law of induction [1], the induced voltage in the secondary coil is:

$$V(t) = -\frac{d\Phi_m(t)}{dt} \tag{2.4}$$

where $\Phi_{\mathrm{m}}(t)$ is the total time-varying magnetic flux crossing the second coil, $V(t)$ is the induced voltage in the second coil. The voltage would cause an induced current in the secondary coil. An induced magnetic field is also generated. According to the Faraday's law [1], the polarity of the induced magnetic field is such that it produces a magnetic field opposes the change which produces it. Because the induced voltage and the current are produced at the secondary side, power is successfully transferred to the secondary side. This is the basic working principles of the wireless power transfer or the inductive coupling.

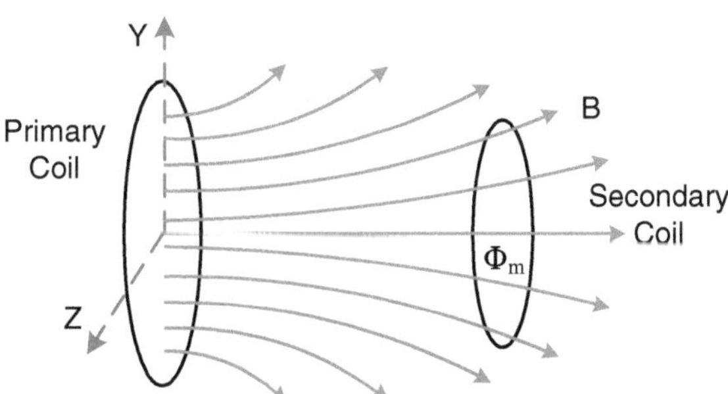

Fig. 2.2 Magnetic field crossing the secondary coil

2.1.2 Transformer Basis

Another way to analyze the wireless power transfer is to use the transformer theory [2]. A transformer is a power converter that transfers AC electrical energy through inductive coupling between circuits of the transformer's primary and secondary coils.

As Fig. 2.3 shows, a transformer has a primary coil with N_1 turns and a secondary coil with N_2 turns.

An ideal transformer is the most basic transformer circuit. In an ideal transformer, it assumes that all magnetic flux generated by the primary coil links all the turns of every secondary coil. According to the Faraday's law [1], the voltages of the primary and secondary coils can be expressed by:

$$\begin{cases} V_1(t) = -N_1 \frac{d\Phi_1(t)}{dt} \\ V_2(t) = -N_2 \frac{d\Phi_2(t)}{dt} \end{cases} \tag{2.5}$$

where N_1 and N_2 are the turns of the primary and secondary coils. Due to the transformer is ideal, the magnetic flux $\Phi_1(t)$ crossing the primary side exactly equals to the magnetic flux $\Phi_2(t)$ crossing the secondary side.

$$\Phi_1(t) = \Phi_2(t) \tag{2.6}$$

Consequently, the voltage equation for the ideal transformer is given by:

$$V_2(t) = \frac{N_2}{N_1} V_1(t) \tag{2.7}$$

Apparently, the voltages rate is only proportional to the rate of turns between the primary and the secondary sides.

In the real wireless power transfer, the transfer medium is air or human tissue. Therefore, magnetic flux loss exists during the transfer. We next analyze imperfect transformers as shown in the Fig. 2.4. Since the primary and the secondary sides have no physical contact, it can be used in the wireless power transfer.

For imperfect transformers, the concepts of the self-inductance, mutual-inductance, and coupling factor are needed for analysis. The self-inductance L of

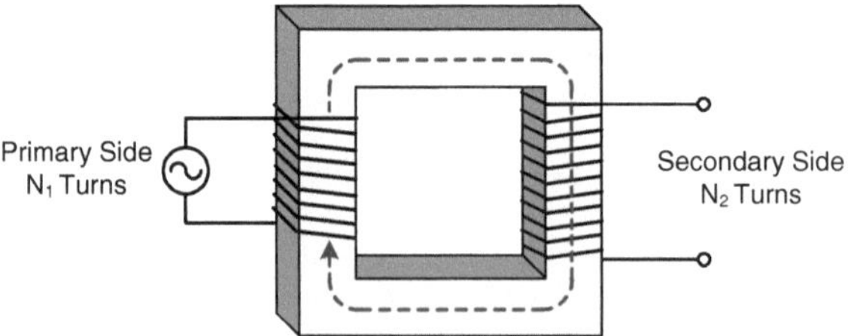

Fig. 2.3 Ideal transformer

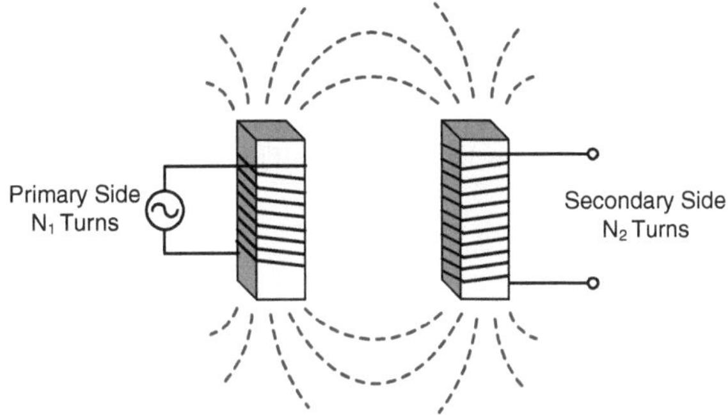

Fig. 2.4 Non-ideal transformer

the coil is defined by dividing the total generated magnetic flux by the current in the coil as follows:

$$L = \frac{N\Phi}{I} \tag{2.8}$$

where N is the coil's turns, I is the current in the coil, and Φ is the magnetic flux crossing each turn of the coil. By substituting the above equation to the Faraday's law [1], the following equation is given as:

$$V = L\frac{dI}{dt} \tag{2.9}$$

where V is the induced voltage in the coil, L is the self-inductance of the coil, I is the time-vary current in the coil. Suppose there is another coil in space and the magnetic field generated by the first coil crosses the second coil. Accordingly, the mutual inductance is defined by the following equations:

$$\begin{cases} M_{21} = \dfrac{N_2\Phi_{21}}{I_1} \\ M_{12} = \dfrac{N_1\Phi_{12}}{I_2} \end{cases} \tag{2.10}$$

where Φ_{21} is the magnetic flux generated by the first coil and crossing the second coil, and the Φ_{12} the magnetic flux generated by the second coil and crossing the first coil. The following equation can be proved:

$$M = M_{21} = M_{12} \tag{2.11}$$

The coupling factor k can be defined by the self and the mutual inductances as the following equation. It describes the coupling strength between the two coils.

$$M = k\sqrt{L_1L_2} \tag{2.12}$$

2.2 System Modeling and Components

In this section, a systematic model for the wireless power transfer for biomedical applications is to be proposed. Three types of the system components are to be introduced.

2.2.1 System Modeling

A wireless power transfer system for the biomedical applications is usually composed of a power transmitter outside of the human body and a power receiver inside of the body or patched on the human skin. Some complex wireless power transfer systems may employ a middle power relay as the third parts to help delivering power to farer place. Figure 2.5 shows the proposed model with a power transmitter, a power receiver, and a middle power relay.

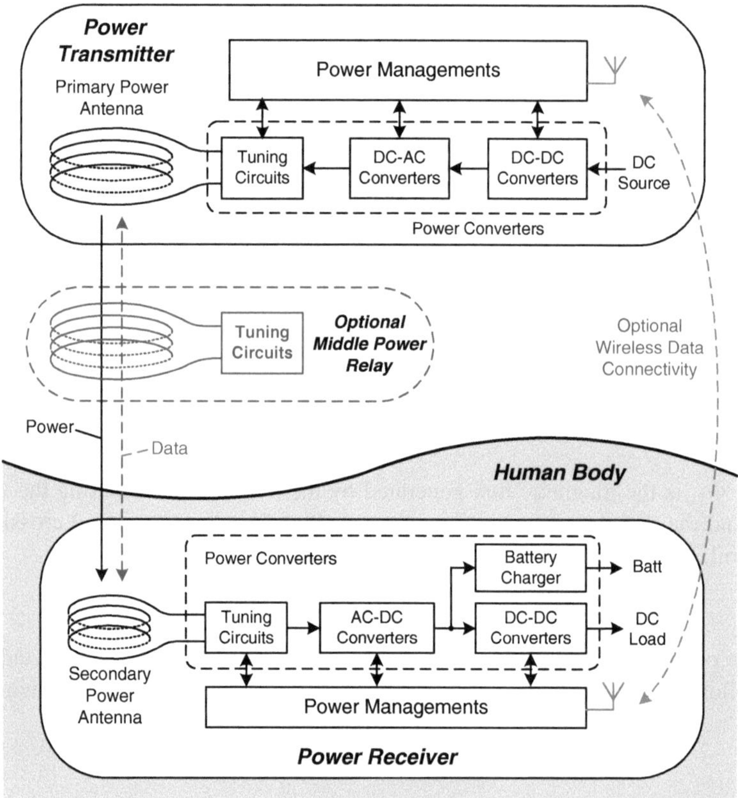

Fig. 2.5 A model of a wireless power transfer system for biomedical microsystems

1. *Power Transmitter*:

The basic function of the power transmitter is to emit wireless power into space. Further functions may include regulating the transferred power level to ensure that recovered power is stable, adaptively changing the operating frequency so that the coupling efficiency is optimized, smartly stopping the transfer when the recharging is over, and so on. The transmitter can be worn on human body, or placed in a room as a location fixed equipment.

Figure 2.5 shows a typical wireless power transmitter. The transmitter is composed of a primary power antenna [3–7], a corresponding tuning circuit [8], a DC–AC converter [9–11], a DC–DC converter [12, 13], and a power management circuit [14]. The DC energy source of the transmitter can be a group of batteries or a city power adapter. The DC–DC converter is responsible to convert the DC energy source to another designed DC voltage for the DC–AC converters and the power management circuits. The DC–DC converter can be fulfilled by linear regulators [15] or switch-mode DC–DC converters [13]. The DC–AC converter in the Fig. 2.5 is the core of the wireless power transmitter. It converts DC energy to AC energy at a specified operating frequency. The DC–AC converter is also called the power inverter [9–11]. The power efficiency of the transmitting side is essentially decided by the efficiency of the inverter. Tuning circuit is necessary to be adopted to allow the antenna to resonant at the designed operating frequency. The power management circuit is used to control the whole transmitter. Typical task includes adjusting the operating frequency and the transferred power level. It may also have wireless data connectivity with the power receiver. The data connectivity can be used to send command to the receiver. It can be also designed to feedback power information from the receiver to the transmitter, so a close-loop power control can be set up.

2. *Optional Middle Power Relay*:

The function of the middle power relay is to resonate at the operating frequency and help to deliver wireless power to farer place. The middle power relay is not a necessary part. It is usually adopted when the power transmitter is not placed near the human body but positioned over a distance from the human body. In this case, the middle power relay worn on the human can be very useful to help the transfer. Typical middle power relays are realized by the LC resonators [16] or the helical antennas [17].

3. *Power Receiver*:

The function of the receiver is to receive the wireless power and provide stable energy to the load or charge the energy into a battery. Further functions of the receiver may include feeding back power information to transmitter so that the transfer power can be regulated, smartly changing the power consumption of the load so that it can adapt to the recovered power, and so on. The receiver may be implanted into the human body, swallowed into the digestive track, or patched on the skin.

As Fig. 2.5 shows, the receiver is composed of a secondary power antenna, a corresponding tuning circuit, an AC–DC converter [18–21], a DC–DC converter [13], a battery charger [22–25], and a power management unit [14]. The tuning circuit makes sure the secondary antenna resonate at the operating frequency. The AC–DC converter converts the AC energy back to DC power. The AC–DC converter can be also called rectifier [18–21], which maybe is a full-wave rectifier [26], a half-wave rectifier [27], a switch-mode rectifier [20], or any other type of rectifiers. The rectifier is definitely the most important circuit in the power receiver. The power receiving efficiency is essentially determined by the rectification efficiency. Because the output of the rectifier is unregulated DC power, DC–DC converter is adopted in the receiving side to provide regulated and stable DC power for the load. The load is actually an application circuit or a rechargeable battery. The power management circuit in the Fig. 2.5 is employed to control the whole receiver. It may be used to monitor the recovered power level and send the power information back to the transmitter using wireless data connectivity.

According to the introduction above, the wireless power transfer for biomedical applications is composed of many modules. Actually, these modules can be classified into three types of system components. They are the power antennas, the power converters, and the power management circuits. Next, we give a brief introduction to the three components. The three components correspond to Chaps. 3, 4, and 5 respectively.

2.2.2 Component Type I: Power Antennas

The power antennas are widely adopted in the wireless power transmitters, receivers, and middle power relays. As shown in the Fig. 2.5, the power antenna in the transmitter excite near-field magnetic field in space. It converts electrical power in circuit to magnetic power in space. The antenna in the middle power relay resonates with the magnetic field, so the magnetic flux density nearby is enhanced. At last, the power antenna in the receiver produces induced voltage and current, which recovers the magnetic power back to the electrical power in circuit.

The power antennas actually play an essential role in the wireless power transfer. The total power transfer efficiency is determined by two efficiencies. One is the coupling efficiency in space. The other is power conversion efficiency in circuits. Generally speaking, the coupling efficiency in the space is much lower than the power conversion efficiency in the circuit. Besides the efficiency, there are other considerations for the power antennas, for example the power level, the size limitation, the quality factor, the medium loss, and so on. Accordingly, the design of power antennas is a quite a comprehensive task.

Typically, the power antennas are loop coils. Tuning capacitor would be adopted to form a LC resonant circuit. At present, many biomedical microsystems are employing this type of power antennas for its decent coupling performance over short distance [4, 19, 20]. To further improve the coupling performance, kinds of new antenna techniques and transfer structure were proposed in recent years. Some of these antennas emphasize to enhance the power efficiency [7, 28], while some emphasize to reduce antenna size [8], realize full directionality [29], and so on.

The details of the power antennas are to be introduced in Chap. 3.

2.2.3 Component Type II: Power Converters

Both the wireless power transmitter and the receiver need power converters. As shown in the Fig. 2.5, the power converters in the transmitter include at least a DC–DC converter and a DC–AC converter. At the receiving side, the power converters include at least an AC–DC converter and a DC–DC converter. The battery recharging circuit [22–25] in the receiver is also a kind of power converter. All these converters may needs other auxiliary circuits, like a Bandgap [30, 31] to provide a reference voltage, a bias circuit [32] to provide a bias current, an oscillator [33, 34] to provide a clock, and so on.

The power converters are significantly important in the system. Researchers try to design power converters with higher power conversion efficiency. For example, conventional rectifiers use diodes or Schottky diodes to recover AC power. However, the dropout voltages of diodes or Schottky diodes are too large for low-voltage low-power applications. Accordingly, many efficiency-enhanced rectifiers are proposed for the wireless power transfers, like the comparator based rectifier [20, 35–37], the rectifier with ZCP prediction [38], and so on. There are so many other state-of-the-art power converters.

The details of the power converters are to be described in Chap. 4.

2.2.4 Component Type III: Power Management

For relatively simple biomedical applications, the power management circuit may not be adopted. For those complex applications, both the power transmitter and the receiver may employ power management. The power management is a relatively vague concept. It may have many functions and it can be viewed as a central controller for the transmitter or the receiver. As shown in the Fig. 2.5, the power management circuits have connections to all power converters. Typically functions include determining the operating frequency, determining the transmitting power level, adjusting the power consumption of the load, and so on. The power management may also act like a power monitor in the receiver. When the voltage, current, or temperature is out of normal range, the power management could trigger an alarm. It can also be used for other special functions.

The details of the power management are to be illustrated in Chap. 5.

2.3 Design Challenges

As introduced above, the wireless power transfer is a promising way to offer energy for biomedical microsystems. However, the transfer still suffers from some problems including the unsatisfied power efficiency, the limited transfer distance, the unpredictable reliability, the thermal issues and other problems. Therefore, this section investigates in-depth the design challenges of the wireless power transfer in biomedical applications. First, we introduce the overall systematic design challenges. Then, both the design challenges at the transmitter and the receiver sides are to be discussed respectively.

2.3.1 Systematic Challenges

Generally speaking, there are five systematic design challenges. They are the transmitter's and receiver's sizes, the transfer distance, the transfer medium, the lateral and angle misalignments, and the required power level.

- Systematic Challenge I: Transmitter's and Receiver's Sizes

In typical biomedical applications, the wireless power transmitter is placed out of human body and the power receiver is located inside of the body or patched on the skin. Accordingly, the receivers usually have much smaller size than the transmitters. For example, the size of a hearing aid [39, 40] is usually only several millimeters, and the size of a pacemaker [41, 42] is limited to several centimeters. On the contrary, because the power transmitter is outside of human body, their size can be much larger. At present, one of the developing trends of the biomedical microsystems is miniaturization. It's a huge challenge for the designers. The decrease of the size would degrade the antenna's quality factor, decrease the coupling factors between the transmitter and the receiver, and force to increase the operating frequency and the energy loss in human tissue.

- Systematic Challenge II: Transfer distance

One of the most critical parameters in the wireless power transfer is the transfer distance. The ratio of the transfer distance and the transmitter's and the receiver's sizes is the key parameter. When the antenna's size is very small and the transfer distance is much longer, the coupling factor would become very small. Transferring power efficiently would become a huge challenge. For instance, the endoscopic capsule [43] is in diameter of around 1 cm, and the required transfer distance is in range from 1 to 10 cm. Consequently, the power transfer efficiency is only 0.3 % [43] in the worst case.

- Systematic Challenge III: Transfer medium

The transfer medium significantly affects the performance of the wireless power transfer. Biomedical applications involve three medium types. They are the air, the human tissue, and the metals. The air has the minimum absorption of electromagnetic field. The human tissue would absorb certain electromagnetic energy according to the operating frequency and the transferred power level. The absorbed energy would cause heat or even get patients hurt. Because some implantable devices use metal shell to protect the internal circuits, the electromagnetic phenomenon of eddy effect [2] appears in the metals. Accordingly, energy is wasted in the metal shell. The antenna in the shell might only receive little energy.

- Systematic Challenge IV: Lateral and angle misalignments

During the use of the wireless power transfer system, there might be lateral and angle misalignments between the power transmitter and the receiver. The definitions of the lateral and angle misalignments are depicted in the Fig. 2.6. The two misalignments cause the degradation of the coupling factor between the transmitter and the receiver. As a result, the transfer efficiency decreases. Making sure the transfer system could work under certain misalignments is very necessary. For some implantable devices like the endoscopy capsule [43], it randomly rotates in the digestive track. Accordingly, special antennas and circuits should be considered.

- Systematic Challenge V: Required Power Level

The required power level is the last systematic design challenge. Satisfying the power demand of the power receiver is the core task of the transmitter. Implantable microsystems like the hearing aid [41, 42], the signal recorder [20], and the nerve stimulator [44] consume tiny energy, typically less than 10 mA. For those low-power applications, how to increase the power transfer efficiency is a design

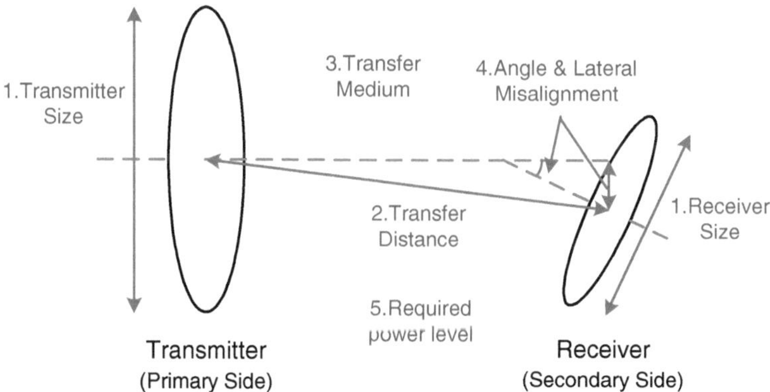

Fig. 2.6 Overall transfer design challenges

challenge because the power efficiency typically decreases when the power level decreases. To the contrary, for those high-power applications, like the artificial heart [45], how to avoid thermal issue at the transmitter side and how to keep the electromagnetic energy absorbed by patients as low as possible are the key challenges.

2.3.2 Challenges at the Transmitter Side

To design a successful wireless power transfer system, it is necessary to understand all detailed challenges at the both transmitter and receiver sides. We firstly investigate the design challenges in the transmitter side. Although some applications have special design challenges, there are five common challenges for the transmitter. They are the peak transfer power, the power dynamic range, the power conversion efficiency, the transmitter size and weight, and the safety issue. Figure 2.7 shows the challenges.

Transmitter challenge I: Peak transfer power
The responsibility of the wireless power transmitter is to satisfy the power demand of the receiver at anytime. For example, a typical wirelessly powered endoscopic capsule consumes the power of 10–30 mW [43]. In the worst case, the transfer efficiency is only around 0.3 % [43]. Consequently, the power transfer is required to offer the peak power of at least 10 W. Although it's quite difficult to know the exact power transfer efficiency before the antennas and circuits are designed, it's still very necessary to predict the peak power for the transmitter before detail designs. It helps to estimate the transmitters' thermal and safety issues.

Transmitter challenge II: Power dynamic range
In some wireless power transfers, the coupling efficiency may change in a wide range. For example, the highest power efficiency of the wirelessly powered capsule endoscopy reaches 5.2 % [43]. However, in the worst case, the efficiency falls to merely 0.3 % [43]. In order to save power and minimize the harmful electromagnetic radiation for the patients, some systems dynamically adjust the

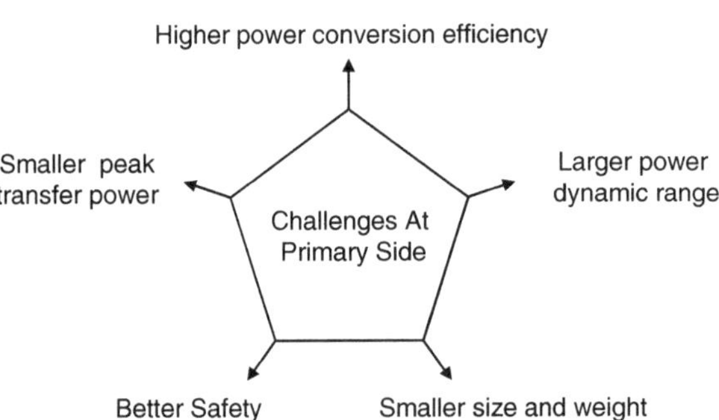

Fig. 2.7 The five design challenges in the transmitter

transmitting power to adapt to the changing coupling efficiency or the changing power consumption of the receiver. As a result, the power transmitter needs to adjust the transferred power level in a broad dynamic range. It's one of the design challenges.

Transmitter challenge III: Peak Conversion Efficiency
The power conversion efficiency of the transmitter is dominant by the DC-AC converters, which are typical realized by Class D or Class E power amplifiers. The Class D amplifier is more commonly used since it's insensitivity to the change of output power or the operating frequency. In the transmitter, the power conversion efficiency is significantly affected by the operating frequency. The efficiency goes down when the frequency increases. Designers have to face the design challenge of trading off between the high operating frequency and the high conversion efficiency. Actually, the selection of the operating frequency has to consider much more factors. The factors include at least the antenna's size, and the specific absorption rate [46], and the power conversion efficiency at the receiver side. It makes the selection of the operating frequency much more complicated. Once the frequency is determined, designers may take the advantage of circuit techniques to improve the power conversion efficiency.

Transmitter challenge IV: Transmitter's Size and Weight
Although the size and weight limitation on the power transmitter is not as critical as the limitation on the power receiver, some applications still have size or weight requirement on the transmitter. For example, the conventional power transmitter for the batteryless capsule endoscopy connects to the city power by using long power cables, which makes patient's movement very inconvenient. To avoid the power cable and allow patients to walk freely, the new transmitter is supposed to use portable batteries as the power source. For the patient's convenience, the transmitter is supposed to be as small and light as possible.

Transmitter challenge V: Better Safety
For the transmitter, two safety issues have to be concerned. One is the thermal problem. The other is the electromagnetic safety issue.

Because the power level at the transmitter side is much higher than the receiver side, the transmitter produces much more heat. The heat is mainly generated by the power transistors in the DC-AC converter and by the conductor resistance of the transmitting antenna. When the power level or the operating frequency increases, the thermal problem becomes serious. Heat sinks and fans can be used. For those high power applications, like the artificial heart [45], the thermal issue at the receiver side should be also concerned.

The other safety issue is the harmful electromagnetic radiation. First, the radiation may interference the normal work of other implanted electronics like the pacemaker [41, 42]. Second, the human expose to electromagnetic radiation may cause risks and hazards. The exposure to electromagnetic fields at frequency above about 100 kHz can lead to significant absorption of energy and temperature increases [47]. Although there are still no authoritative standard dedicated for the electromagnetic radiation caused by wireless power transfer for medical usages, designers could take two famous international electromagnetic standards as references. One is the IEEE Standard C95.1 [48]. The other is the ICNIRP Guidelines [47]. A lower operating frequency may be considered to reduce the heat and the radiation.

Summarizing the above concerns at the transmitter side, the key question is how to build a wireless power transmitter that has large peak power, wide power dynamic range, high power conversion efficiency, small size and light weight, little heat, and little radiation for the patients.

2.3.3 Challenges at the Receiver Side

At the secondary side, there are also five challenges. They are the power demand, the supply voltage, the power conversion efficiency, the power reliability, and the receiver size. Figure 2.8 shows the challenges.

Receiver challenge I: Power Demand
The power demand of the wireless power receivers differs from applications. For example, the nerve stimulator and the capsule endoscopy usually consume the power of 10–30 mW [43, 44]. However, the wirelessly powered artificial heart consumes a lot more power, like up to 15 W [45]. How to meet the peak power demand of the application circuits is always the first concern in the wireless power receiver's design.

Receiver challenge II: Supply voltages
The load of the power receiver is actually applications circuits. They usually have specific requirement on the supply power voltage. For example, if the load is a digital circuit, it may require a power supply voltage like 1.8 or 3.3 V. Because the AC-DC converter in the receiver outputs unregulated DC voltage, DC–DC converters have to be used to generate well-regulated voltage for the loads. A special problem is that the rectified DC voltage varies with the transfer distance and the

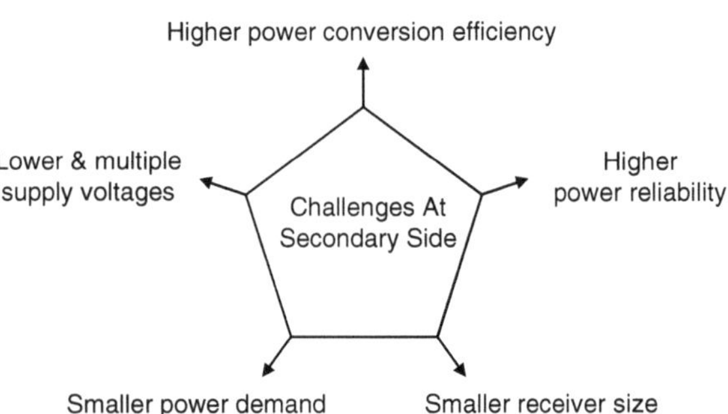

Fig. 2.8 The five design challenges in the receiver

relative orientation between the transmitter and the receiver. As a consequence, the rectifier output is likely lower or higher than the required voltages by the load. Special circuit techniques are needed to solve this problem.

Receiver challenge III: Power conversion efficiency
The power conversion efficiency is definitely one of the most important concerns in the design of the wireless power receiver. According to existing wireless power receiver designs [8, 18–21, 38], the power conversion efficiency unavoidably goes down when the operating frequency increases or the transfer power decreases. As a result, it is a big challenge to design a low-power but high-efficiency power receiver, especially when the operating frequency is very high. For example, the power efficiency of a rectifier working at 900 MHz with output power of only 140 μW is no more than 65 % [8]. To the contrary, to transfer large power at a low operating frequency can be much more energy-efficient, like 93.6 % [38]. Since the power level is determined by the load, the designer's only choice is the operating frequency.

Receiver challenge IV: Power reliability
Comparing to a power cable, the wireless power system is much more unstable. The unstable factors arise because the separate distance between the transmitter and the receiver may suddenly changes, the relative orientation between the transmitter and the receivers may also alters, and the magnetic field may be affected by other unexpected devices. Consequently, the reliability of the wireless power receiver becomes unpredictable for the load. Designers should clarify all unstable factors in systems. For example, how can we ensure the application circuit could successfully start up when the wireless power is transmitted at the very beginning? What if the receiver suddenly rotates? If the power recovered by the receiver falls slowly, what is the response of the application circuits? Would it trigger any error operation? It is a challenge to answer these questions. Power management may be adopted for the promotion of the power reliability.

Receiver challenge V: Receiver's Size
Receiver's size is another problem we have to consider, especially for those ultra small size implantable biomedical devices. For example, researchers have pronounced the mm-sized wirelessly powered system [8]. The receiver's size is determined by the receiving antenna's size and the circuit's size. As to the antenna, it can be significantly decreased by using higher operating frequency. However, as mentioned, the power conversion efficiency goes down when the operating frequency increases. As to the circuit, the circuit's size is actually decided by the passive components employed in the circuits. For example, rectifiers need decoupling capacitors at the output terminal. Linear regulators require decoupling capacitors at input and output terminals. DC–DC converters need off-chip inductors. In most of time, the sizes of these passive components are too large to be integrated into microchip. Consequently, they are the main contributions to the circuit's size. How to design a small size receiver is a challenge.

Summarizing the above concerns at the receiver side, the key question is how to build a wireless power receiver that has wide input voltage, small size, high power reliability, and high power conversion efficiency in the low-power high-frequency condition.

2.4 Electromagnetic Safety

Comparing to regular cable power systems, the wireless power transfer suffers from unique safety issues. One of the concerns is the risk caused by the expose to the electromagnetic radiation. Expose to low-frequency electric and magnetic fields normally results in negligible energy absorption and no measurable temperature rise in the body [47]. However, exposure to electromagnetic fields at frequency above about 100 kHz can lead to significant absorption of energy and temperature increases [47]. Accordingly, understanding and considering the radiation safety issues are very necessary. In this section, we firstly introduce two international electromagnetic safety standards. Secondly, we present several clinical experiences as references for the systematic design.

2.4.1 Safety Standards

Although there are still no authoritative international standard dedicated for the wirelessly power transfer for biomedical systems, there are two international electromagnetic standards for public safety that we could take as references. One is the "IEEE Standard for Safety Levels with Respect to Human Exposure to Electromagnetic Field" [48] from the Institute of Electrical and Electronics Engineers (IEEE). The other is the "ICNIRP Guidelines for Limiting Exposure to Time-Varying Electric, magnetic, and Electromagnetic Fields" [47] from the International Commission on Non-Ionizing Radiation Protection (ICRIRP).

1. IEEE Standard [48]: The Institute of Electrical and Electronics Engineers is founded in 1884 and is the world's largest technical professional society. The international committee on Electromagnetic Safety (ICES) develops standards for the safe use of electromagnetic energy. The ICES is sponsored by IEEE. In 1960, IEEE and U.S. Navy co-sponsored the development of the first US national RF standard (C95.1-1966). Later, ICES published a group of standards including C95.6-2002 "IEEE Stand for Safety Levels with Reset to Human Exposure to Electromagnetic Field, 0–3 kHz" and C95.1 "Standard for Safety Levels with Respect to Human Exposure to Radio Frequency Electromagnetic Fields, 3 kHz–300 GHz". Due to most wireless power transfers work at the operating frequency in range from 100 K to 50 MHz, we introduce the standard of IEEE C95.1 but not the C95.6.

2. ICRIRP Guideline [47]: In 1974, the International Radiation Protection Association (IRPA) formed a working group on non-ionizing radiation. At the IRPA congress in Paris in 1977, this working group became the International Non-Ionizing Radiation Committee (INIRC). In cooperation with Environmental Health Division of the World Health Organization (WHO), the INIRC developed a number of hearth criteria documents. In 1992, a new independent scientific organization,—the International Commission on Non-Ionizing Radiation Protection (ICNIRP) was established as a successor to the INIRC. The functions of the Commission are to investigate the hazards that may be associated with different forms of NIR, develop international guidelines on NIR exposure limits, and deal with all aspects of NIR protection. Its publication "ICRIRP Guidelines for Limiting Exposure to Time-Varying Electric, Magnetic, and Electromagnetic Fields" is right such a document.

Both the IEEE standard and the ICNIRP guideline give the maximum permissible exposure. The IEEE standard covers the frequency in range from 1 Hz to 300 GHz. The ICNIRP guideline applies to frequency in range from 3 kHz to 300 GHz. Both of them cover the typical operating frequency range in the wireless power transfer, which is from 0.1 to 100 MHz. The given permissible exposure is ruled the maximum magnetic flux density and the maximum electrical field strength.

(a) Maximum magnetic flux density:

A magnetic field can be specified in two ways, the magnetic flux density B expressed in Tesla (T), or the magnetic field strength H expressed in ampere per meter (A/m). Figure 2.9 shows the maximum permissible exposure of time-varying magnetic field expressed in Tesla (T) ruled by the two documents.

As Fig. 2.9 shows, the maximum permissible exposure magnetic field decreases when the operating frequency increases. It is noted that there are two environments in each documents. In the ICNIRP guideline, the two environments are called the "occupational" and the "general public". In the IEEE standard, the two environments are the "controlled" and the "uncontrolled". Both the occupational and the controlled environments mean locations where there is exposure that is incurred by persons who are aware of the potential for exposure as a concomitant as a concomitant of employment. Both the general public and the uncontrolled mean where there is the exposure of individuals who have no knowledge or control of their exposure. Accordingly, the maximum permissible expose in the occupational and controlled environments are much higher than the general public and the uncontrolled environments. The detail limits in the two documents in the interesting frequency range for wireless power transfer are summarized in the following Table 2.1.

According to the definitions of the "occupational" or the "controlled" environments, we suggest the maximum permissible exposure caused by the wireless power transfer for biomedical purposes could be larger than the current

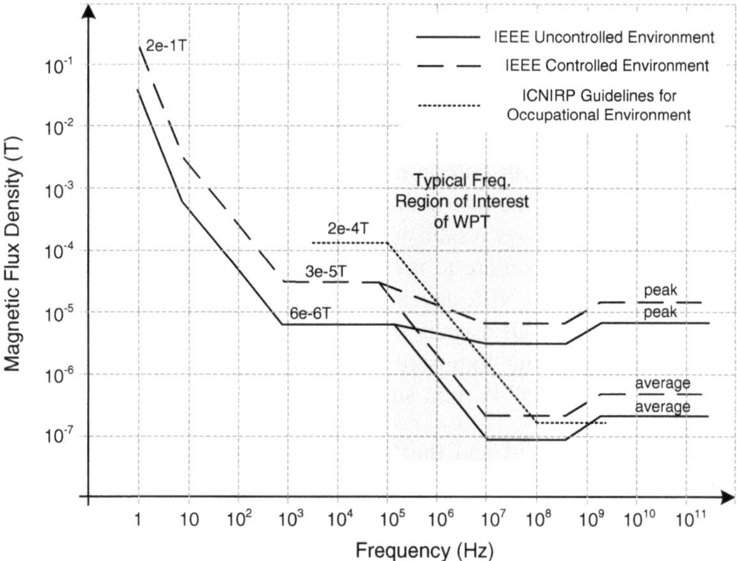

Fig. 2.9 Maximum permissible exposure of time-varying magnetic field suggested by the IEEE standard and the ICNIRP guideline

Table 2.1 Maximum permissible exposure of time-varying magnetic field suggested by the IEEE standard and the ICNIRP guideline

Frequency range f in MHz	IEEE controlled environment (μT)	IEEE uncontrolled environment (μT)	ICNIRP occupational environment (μT)	ICNIRP general public environment (μT)
0.1–0.15	20.4/f	20.4/f	2.0/f	6.25
0.15–1				0.92/f
1–10				
10–30			0.2	0.092
30–100		$199/(f^{1.668})$		

Note The expose values in terms of magnetic field strengths are the mean values obtained by spatially averaging the squares of the fields over an area equivalent to the vertical cross section of the human body

permissible exposure for "occupational" environments. It is because the two documents deal with generally healthy people. But, the wireless power transfer for biomedical applications deals with people in risk of diseases or even death. The wirelessly powered systems may save the patient's life.

Can the wireless power transfer comply with the permissible magnetic flux density? The magnetic field generated by typical low-power transfers like the percutaneous transfer [8] could comply with the documents. However, for those high-power systems that transfer dozens of watts, the magnetic flux density may reach 1.25–10 μT [3] when the limitation is about 2 μT at 10 MHz in the IEEE standard. As a reference, the Magnetic Resonance Imaging (MRI) produces a much larger magnetic flux density in range from 500,000 to 1500,000 μT in strength [49].

(b) Maximum Electric Field Strength:

Not only the magnetic field strength is ruled, but also the maximum electric field strength is limited. An electric field, *E*, experts forces on an electric charge and is expressed in volt per meter (V/m). The Fig. 2.10 shows the maximum permissible exposure of time-varying electric field expressed in volt per meter (V/m) regulated by the two documents. It's clear that the maximum permissible exposure electric field decreases when the operating frequency increases. The detail limits in the two documents in the frequency range for the wireless power transfer are summarized in the Table 2.2.

Can the wireless power transfer comply with the permissible electric field? Actually, for those high-power wireless power transfers, they may easily exceed the limitation. For instance, to transfer 60 W over 2 m, previous research produced an electrical field in range from 210 V/m to 1.4 kV/m in the frequency of 9.9 MHz [3]. The suggested maximal electric field is only 61 V/m for the occupational environment in the ICNIRP guideline.

For both electric and magnetic fields, the maximum peak permissible value and the maximum average permissible value are different. The average values are

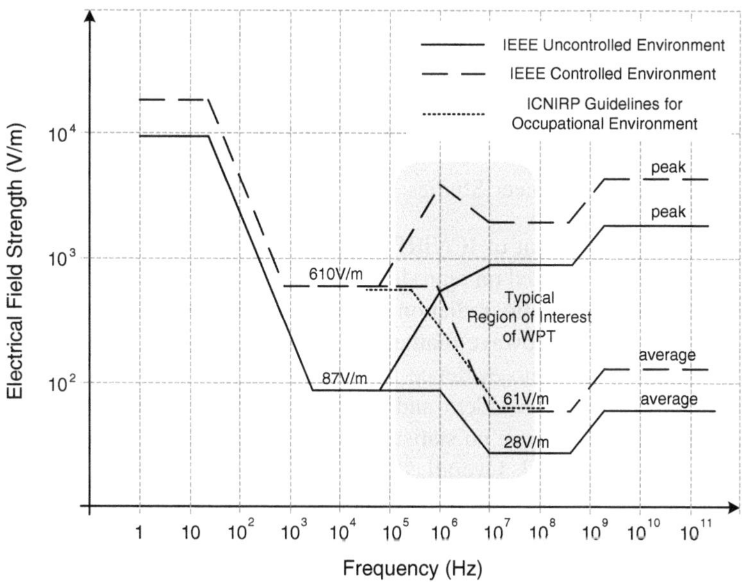

Fig. 2.10 Maximum permissible exposure of time-varying electrical field suggested by the IEEE standard and the ICNIRP guideline

Table 2.2 Maximum permissible exposure of time-varying electric field suggested by the IEEE Standard and the ICNIRP guideline

Frequency range f in MHz	IEEE controlled environment (V/m)	IEEE uncontrolled environment (V/m)	ICNIRP occupational environment (V/m)	ICNIRP general public environment (V/m)
0.1–1.34	614	614	610	87
1.34–3		823/f	610/f	$87/f^{0.5}$
3–10	1842/f			
10–30		27.5	61	28
30–100	61.4			

Note The expose values in terms of electric field strengths are the mean values obtained by spatially averaging the squares of the fields over an area equivalent to the vertical cross section of the human body

usually measured based on any 6-min period [47], and the peak value is suggested about 20 times of the average value.

2.4.2 Clinical Experiences

Both the IEEE standard and the ICNIRP guideline are based on clinical literatures and experiences on the biological effects and potential health effects of electromagnetic fields. Learning these clinical literatures and experiences helps us understanding the effects caused by the wireless power transfer. Meanwhile, because some wireless power transfers for the biomedical microsystems may exceed the suggested maximum permissible expose in two above documents, it is necessary for us to know the consequence of over-exposing. According to the ICNIRP guidelines [47], there are following clinical experiences.

(a) Reproductive and Cancer Studies:

According to the statement of ICNIRP guidelines [47], only a limited number of studies have been conducted on reproductive effects and cancer risks in individuals exposed to electromagnetic radiation. For example, two extensive studies on woman treated with microwave diathermy to relieve the pain of uterine contractions during labor found no evidence for adverse effects on the fetus [50]. In the studies of female plastic welders and physiotherapists working with shortwave diathermy device, there were no statistically significant effects on rate of abortion or fetal malformation [51]. Overall, the studies on reproductive outcomes suffer from poor assessment of exposure and small numbers of subjects. It is difficult to draw firm conclusions on reproductive risk without more assessment [47]. The cancer studies also have little quantitative assessment. Overall speaking, the results of the small number of reproductive and cancer studies are inconclusive.

Table 2.3 A summary of existing experimented exposes in terms of SAR and their corresponding clinical adverse effects

SAR (W/kg)	Experiment subject	Corresponding effects
100–140	Rabbit	Cataracts in eye [54]
7–15	Mice	Slight heating [56]
4	Volunteer	Increased body core temperature of less than 1 °C [55]
1–3	Rat and Monkey	Decreased task performance [57, 58]

Note The safety SAR limited by ICNIRP is 0.4 W/kg, which is much lower than the values of the SAR in these experiments

(b) Volunteer and Animal Studies:

According to previous studies [52], as the frequency increases from 100 kHz to 10 MHz, the dominant effect of exposure to a high-intensity electromagnetic field changes from nerve and muscle stimulation to heating. At 100 kHz the primary sensation was one of nerve tingling, while at 10 MHz it was one of warmth on the skin. In this frequency range, therefore, basic health protection criteria should be such as to avoid stimulation of excitable tissues and hating effects. For example, there have been several studies of thermoregulatory response of testing volunteers exposed to electric magnetic field in magnetic resonance imaging systems [47]. In general, these have demonstrated that exposure for up to 30 min, under conditions in which whole-body Specific Energy Absorption (SAR) was less than 4 W/kg, caused an increase in the body core temperature of less than 1 °C [47]. There are numerous reports on the behavioral and physiological response of laboratory animals, including rodents, dogs, and non-human primates [47]. Previous studies [53] shows exposure of laboratory animals to electric magnetic field producing absorption in excess of approximately 4 W/kg has revealed a characteristic pattern of thermoregulatory response in which body temperature initially rises and then stabilizes following the activation of thermoregulatory mechanisms. Meanwhile, decreased task performance by rats and monkeys has been observed at SAR values in range 1–3 W/kg. Microwave exposure of 2–3 h duration has produced cataracts in rabbits' eye at SAR values from 100 to 140 W/kg, which produced lenticular temperature of 41–43 °C [54].

(c) Summary and Conclusions:

We summarized the existing experience in Table 2.3. It sums up the clinical experiences when the SAR is increased from 1 to 140 W/kg.

To conclude, the exposure for approximately 30 min to electric magnetic field producing a whole-body SAR between 1 and 4 W/Kg results in body temperature increase of less than 1 °C [55]. Even the most sensitive tissues can handle the exposure greater than 4 W/kg [47], which can be used as the safety threshold for wireless power transfer systems for biomedical microsystems.

References

1. Thidé, B. (2004). *Electromagnetic field theory*. Uppsala: Upsilon Books.
2. Winders, J. (2002). *Power transformers principles and applications*. New York: Marcel Dekker Inc.
3. Kurs, A., Karalis, A., Moffatt, R., Joannopoulos, J. D., et al. (2007). Wireless power transfer via strongly coupled magnetic resonances. *Science, 317*(5834), 83–86.
4. O'Handley, R. C., Huang, J. K., Bono, D. C., et al. (2008). Improved wireless, transcutaneous power transmission for in vivo applications. *Sensors Journal, IEEE, 8*(1), 57–62.
5. Casanova, J. J., Low, Z. N., & Lin, J. (2009). A loosely coupled planar wireless power system for multiple receivers. *IEEE Transactions on Industrial Electronics, 56*(8), 3060–3068.
6. Cannon, B. L., Hoburg, J. F., Stancil, D. D., et al. (2009). Magnetic resonant coupling as a potential means for wireless power transfer to multiple small receivers. *IEEE Transactions on Power Electronics, 24*(7), 1819–1825.
7. Imura, T., Okabe, H., Uchida T., & et al. (2009) Study on open and short end helical antennas with capacitor in series of wireless power transfer using magnetic resonant couplings. *IECON* (pp. 3848–3853).
8. O'Driscoll, S., Poon, A., & Meng, T. H. (2009). A mm-sized implantable power receiver with adaptive link compensation. *ISSCC* (pp. 294–295).
9. Casanova, J. J., Low, Z. N., & Lin, J. (2009). Design and optimization of a class-e amplifier for a loosely coupled planar wireless power system. *IEEE Transactions on Circuits and Systems II: Express Briefs, 56*(11), 830–834.
10. Schwarzer, U., & De Doncker, R. W. (2001). Power losses of IGBTs in an inverter prototype for high frequency inductive heating applications. *IECON, 2*, 793–798.
11. Saitou, M., & Shimizu, T. (2000). A novel strategy of the high power PWM inverter with the series active filter. *ISIE* (vol. 1, pp. 67–72).
12. KuanC.-W., & Lin, H.-C. (2012). Near-independently regulated 5-output single-inductor DC–DC buck converter delivering 1.2 W/mm^2 in 65 nm CMOS. *ISSCC* (pp. 274–276).
13. Joyce,K., Yogesh, R., Naveen, V., & et al. (2008) A 65 nm Sub-VT microcontroller with integrated SRAM and switched-capacitor DC–DC converter. *ISSCC* (pp. 318–616).
14. Jiang, W., Yu, F., Wu, Y., & et al. (2012). *Wireless power management of small scale distributed hydrogen harvesting system with full digital control*. In: International Conference on Advanced Mechatronic Systems (pp. 149–153).
15. Guochen, A., & Zhanyou, S. (2007) Programmable voltage regulator design based on digitally controlled potentiometer. *ICEM* (pp. 1–453).
16. RamRakhyani, A. K., Mirabbasi, S., & Chiao, M. (2011). Design and optimization of resonance-based efficient wireless power delivery systems for biomedical implants. *IEEE Transactions on Biomedical Circuits and Systems, 5*(1), 48–63.
17. Zhang, F., Hackworth, S. A., Fu, W., et al. (2011). Relay effect of wireless power transfer using strongly coupled magnetic resonances. *IEEE Transactions on Magnetics, 47*(5), 1478–1481.
18. Ghovanloo, M., & Najafi, K. (2004). Fully integrated wideband high-current rectifiers for inductively powered devices. *Solid-State Circuits, IEEE Journal of, 39*(11), 1976–1984.
19. Yoo, J., Yan, L., Lee, S., Kim, Y., et al. (2010) A 5.2 mW self-configured wearable body sensor network controller and a 12 uW 54.9 % efficiency wirelessly powered sensor for continuous health monitoring system. *ISSCC* (pp. 178–188).
20. Lee, S. B., Lee, H.-M., Kiani, M., et al. (2010). An inductively powered scalable 32-channel wireless neural recording system-on-a-chip for neuroscience applications. *IEEE Transactions on Biomedical Circuits and Systems, 4*(6), 360–371.
21. Nakamoto, H., Yamazaki, D., Yamamoto, T., & et al. (2006). A passive UHF RFID tag LSI with 36.6 % efficiency CMOS-only rectifier and current-mode demodulator in 0.35 μm FeRAM technology. *ISSCC* (pp. 1201–1210).

22. Yao, Y., Zhang, H., & Geng, Z. (2011). Wireless charger prototype based on strong coupled magnetic resonance. *EMEIT* (pp. 2252–2254).
23. Ning, L., Xiao, Y., & Ning, Z. (2011). Design of transcutaneous coupling wireless charger. *ICCSE* (pp. 41–46).
24. Dai, D., & Liu, J. (2012). Human powered wireless charger for low-power mobile electronic devices. *IEEE Transactions on Consumer Electronics, 58*(3), 767–774.
25. Jiang, H., Brazis, P., Tabaddor, M., & Bablo, J. (2012). Safety considerations of wireless charger for electric vehicles—A review paper. *ISPCE* (pp. 1–6).
26. Karthikeyan, L., & Amrutur, B. (2012). Signal-powered low-drop-diode equivalent circuit for full-wave bridge rectifier. *IEEE Transactions on Power Electronics, 27*(10), 4192–4201.
27. Heljo, P. S., Li, M., Lilja, K. E., et al. (2013). Printed half-wave and full-wave rectifier circuits based on organic diodes. *IEEE Transactions on Electron Devices, 60*(2), 870–874.
28. Zhang, F., Liu, X., Hackworth, S. A., & et al. (2009). In vitro and in vivo studies on wireless powering of medical sensors and implantable devices. *LiSSA* (pp. 84–87).
29. Lenaerts, B., & Puers, R. (2006). An omnidirectional transcutaneous power link for capsule endoscopy. *BSN.*
30. Leung, C. Y., Leung, K. N., & Mok, P. K. (2004). Design of a 1.5-V high-order curvature-compensated CMOS bandgap reference. *ISCAS.*
31. Lee, E. K. (2010). Low voltage CMOS bandgap references with temperature compensated reference current output. *ISCAS* (pp. 1643–1646).
32. Minch, B. A. (2002). A low-voltage MOS cascode bias circuit for all current levels. *ISCAS* (pp. 619-622).
33. Lim, J., Lee, K., & Cho, K. (2010). Ultra low power RC oscillator for system wake-up using highly precise auto-calibration technique. *ESSCIRC* (pp. 274–277).
34. Lasanen, K., Raisanen-Ruotsalainen, E., & Kostamovaara, J. (2002). A 1-V, self adjusting, 5-MHz CMOS RC-oscillator. *ISCAS* (pp. 377–380).
35. Sun, Y., Jeong, C., Han, S., & Lee, S. (2001). A high speed comparator based active rectifier for wireless power transfer systems. *IMWS-IRFPT* (pp. 1–2).
36. Cha, H.-K., Park, W.-T., & Je, M. (2012). A CMOS Rectifier With a Cross-Coupled Latched Comparator for Wireless Power Transfer in Biomedical Applications. *IEEE Transactions on Circuits and Systems II: Express Briefs, 59*(7), 409–413.
37. Guo, S., & Lee, H. (2007). An efficiency-enhanced integrated CMOS rectifier with comparator-controlled switches for transcutaneous powered implants. *CICC* (pp. 385–388).
38. Sun, T. J., Xie, X., Li,G., Gu, Y., Li, X., & Wang, Z. (2011). An omnidirectional wireless power receiving IC with 93.6 % efficiency CMOS rectifier and Skipping Booster for implantable bio-microsystems. *A-SSCC* (pp. 185–188).
39. Serra-Graells, F., Gomez, L., & Huertas, J. L. (2004). A true-1-V 300-μW CMOS-subthreshold log-domain hearing-aid-on-chip. *IEEE Journal of Solid-State Circuits, 39*(8), 1271–1281.
40. Qiao, P., Corporaal, H., & Lindwer, M. (2011). A 0.964 mW digital hearing aid system. *DATE* (pp. 1–4).
41. Wong, L. S., Hossain, S., Ta, A., Edvinsson, J., Rivas, D. H., & Naas, H. (2004). A very low-power CMOS mixed-signal IC for implantable pacemaker applications. *IEEE Journal of Solid-State Circuits, 39*(12), 2446–2456.
42. Lee, S.-Y., Su, M. Y., Liang, M.-C., et al. (2011). A programmable implantable microstimulator SoC with wireless telemetry: Application in closed-loop endocardial stimulation for cardiac pacemaker. *IEEE Transactions on Biomedical Circuits and Systems, 5*(6), 511–522.
43. Sun, T., Xie, X., Li, G., et al. (2012). A two-hop wireless power transfer system with an efficiency-enhanced power receiver for motion-free capsule endoscopy inspection. *IEEE Transactions on Biomedical Engineering, 59*(11), 3247–3254.
44. Chiu, H.-W., Lin, M.-L., Lin, C.-W., et al. (2010). Pain Control on demand based on pulsed radio-frequency stimulation of the dorsal root ganglion using a batteryless implantable CMOS SoC. *IEEE Transactions on Biomedical Circuits and Systems, 4*(6), 350–359.

45. Si, P., Hu, A. P., Malpas, S., et al. (2008). A frequency control method for regulating wireless power to implantable devices. *IEEE Transactions on Biomedical Circuits and Systems, 2*(1), 22–29.
46. Shiba, K., Nagato, T., Tsuji, T., et al. (2008). Energy transmission transformer for a wireless capsule endoscope: Analysis of specific absorption rate and current density in biological tissue. *IEEE Transactions on Biomedical Engineering, 55*(7), 1864–1871.
47. *ICNIRP Guidelines For Limiting Exposure To Time-Varying Electric, Magnetic and Electromagnetic Field.* International Commission on Non-Ionizing Radiation Protection, 1998.
48. IEEE Standard for Safety Levels With Respect to Human Exposure to Radio Frequency Electromagnetic Fields, 3 kHz–300 GHz. *IEEE Std C95.*
49. Morrow, G. (2000). Progress in MRI magnets. *IEEE Transactions on Applied Superconductivity, 10*(1), 744–751.
50. Daels, J. (1973). Microwave heating of uterine wall during parturition. *Obstetrics and Gynecology, 42*, 76–79.
51. Kallen, B., Malmquist, G., & Moritz, U. (1982). Delivery outcome among physiotherapists in sweden 0 is nonionizing radiation a fetal hazard. *Archives of Environmental Health, 37*, 81–85.
52. Chatterjee, I., Wu, D., & Gandhi, O. P. (1986). Human body impedance and threshold currents for perception and pain for contact hazard analysis in the VLF-MF band. *IEEE Transactions on Biomedical Engineering, 5*, 486–494.
53. Hand, J. W. (1984). Biological effects and dosimetry of nonionizing radiation. *International Journal of Radiation Biology, 45*(2), 197–198.
54. Guy, A. W., Lin, J. C., Kramar, P. O., & Emery, A. F. (1975). Effect of 2450-MHz radiation on the rabbit eye. *IEEE Transactions on Microwave Theory and Techniques, 23*(6), 492–498.
55. Shellock, F. G., & Crues, J. V. (1987). Temperature, heart rate, and blood pressure changes associated with clinical MR imaging at 1.5 T. *Radiology, 163*(1), 259–262.
56. Repacholi, M. H., Basten, A., Gebski, V., et al. (1997). Lymphomas in Eμ-Pim1 transgenic mice exposed to pulsed 900 MHz electromagnetic fields. *Radiation Research, 147*(5), 631–640.
57. Stern, S., Margolin, L., Weiss, B., et al. (1979). Microwaves: Effect on thermoregulatory behavior in rats. *Science, 206*(4423), 1198–1201.
58. Adair, E. R., & Adams, B. W. (1980). Microwaves modify thermoregulatory behavior in squirrel monkey. *Bioelectromagnetics, 1*(1), 1–20.

Chapter 3
Wireless Power Antennas

Abstract The wireless power antenna plays an important role in the wireless power transfer. To reduce the system size, realize the full directionality, improve the coupling efficiency and other parameters, various kinds of power antenna techniques have been proposed in recent years. Accordingly, these techniques constitute many wireless power transfer structures. In this chapter, we firstly give an overview to the power antenna techniques in the wireless power transfer, introduce their classification and detailed catalogs, and discuss their design considerations. Second, four kinds of transfer structures are presented with details. They are the LC-pair, the Multiple-resonators, the Quad-loops, and the Helix-derivatives. Both basic design methods and advanced optimizations are to be proposed or explained. Although the transfer structures share the same physical principles, they have entirely different performances and application conditions.

3.1 Introduction

A wireless power transfer system is mainly made of antennas and circuits. The antenna actually plays an important role in a wireless power transfer system. In the chapter, we introduce the antenna techniques of the wireless power transfer. Typically speaking, the power conversion efficiency of power circuits is in range of 30–80 % [1–4]. However, the coupling efficiency of antennas varies in a much larger dynamic range, like less than 0.1 % [5] to over 40 % [6–8]. Accordingly, the coupling performance between primary and secondary antenna significantly affects the overall system performances. To improve the performances including the transfer distance, coupling efficiency, antenna size, full directionality, and other performances, various antenna techniques were proposed in recent years. We first take an overview of the power antennas in the wireless power transfer for biomedical microsystems.

T. Sun et al., *Wireless Power Transfer for Medical Microsystems*,
DOI: 10.1007/978-1-4614-7702-0_3,

3.1.1 Power Antenna Overview

An antenna is an electrical device which converts electric power into radio waves. Antennas can be used to transfer data or power. In the wireless power transfer, power antennas are required by transmitters and receivers to couple electrical power in circuit to the power in the electromagnetic field.

The most well-known power antenna for biomedical applications is the magnetic inductive coil. A coil is formed when a conductor, usually a copper wire, is wound around a core to create an inductor. The core might be an air core or a ferrite core, and the coil may have multiple turns. Coils are often coated with varnish or wrapped with insulating tape to provide insulation. When a flow of current passes through the coil, it generates heat. Besides heat, because the coil has inductance, it generates magnetic field. Accordingly, the coil couples electrical power in the circuit to the power in the magnetic field. When there is another coil in space, the magnetic energy can be coupled back to electrical power in the second coil. Because coil is inductive, capacitor is usually adopted to make the coil resonant at designed frequency. A resonant coil improves the power transfer efficiency.

Power antennas can be classified to omnidirectional and directional power antennas. For simple biomedical applications, like percutaneous power transfers [9], directional power antennas are typically adopted. Directional power antennas are intended to transmit or receive in particular direction. The simplest loop coil is right a directional power antenna, which transfers energy along its axial direction. To the contrary, the omnidirectional power antennas receive or transmit power in all directions. They are typically employed when the relative position between the power transmitter and the receiver is unknown or changes frequently. In the wireless power transfer, the omnidirectional power antenna is often formed by three orthogonal coils.

To improve the power transfer efficiency, matching the impedance of the power antenna to the impedance of power converters in the circuits is critical. In most of radio systems, the source and load impedances agree on 50 Ohm. However, in wireless power transfer systems, especially in the low-frequency systems, the impedances are designed to be other values. Forcing all power transmitters and receivers to match the regular impedance of 50 Ohm is unnecessary and is a waste of efficiency. For example, in a power transmitter, a primary power antenna connects to a large-power and low-frequency inverter. The source impedance of the inverter is typically much lower than 50 Ohms, like only 0.5 Ohm. Meanwhile, the primary antenna uses a heavy copper wire to reduce its conductor resistance. As a result, the inverter's and coil's impedance will be matched at a value that is much less than 50 Ohm.

Unlike the regular data antenna in radio that operates in frequency range from 100 MHz to over GHz, the power antenna in biomedical systems typically works under relatively lower frequency, like 100 kHz to 50 MHz. Meanwhile, the power transfer distance is usually no more than 1 m. As a result, typical power antennas work in the near-field range. Some conventional parameters and analyze methods for data antennas, like the gain [10], radiation pattern [10], and polarization [10] cannot be applied to the power antennas. Instead, designers pay more attentions to parameters like the inductance [11], self resonant frequency [11], quality factor [11], internal resistance [11], coupling factor [11], and so on. By improving these parameters, researchers have developed many power antennas for biomedical microsystems. All these parameters and the recent antenna techniques will be introduced in this chapter.

3.1.2 Design Considerations

For power antennas in the wireless power transfer, there are five common considerations. They are the power level, the antenna size and weight, the quality factor, the coupling factor, and the operating frequency. They are showed in Fig. 3.1. It is very necessary to understand these design considerations.

1. Power level: The power level describes how much power the antenna is dealing with. For the power transmitting antenna at the primary side, it is called the emitted power. And for the power receiving antenna at the secondary side, it is called the received power. Of course, the emitted power is higher than the received power. Take the batteryless capsule endoscopy as an example, the emitted power is up to 8 W [12] and the received power is around 100 mW [12].

For those high-power systems, the real challenge is apparently at the primary side. To deal with relatively high power at the primary side, designers have to use heavy wires for large current in the coil. Even though, the thermal issue is still a problem. Fans and heat sink may be employed [12, 13]. For those ultra high-power systems, like the wirelessly powered artificial heart with the consumption of 30 W [14], thermal issue at the secondary coil becomes another problem. Furthermore,

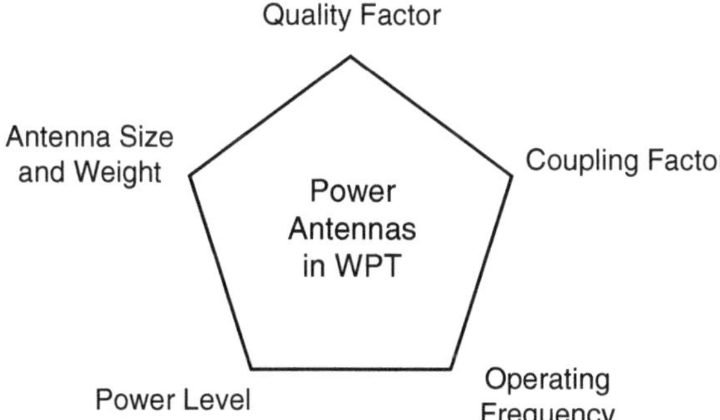

Fig. 3.1 Five design considerations of the power antennas

designing high-power antennas usually requires lower operating frequency. The determination of the frequency involves trading off with other systematic considerations.

2. Antenna size and weight: Typically, biomedical microsystems have very strict requirements on the size and weight. Accordingly, there are strict limitations on the secondary power antenna. For example, the size of a neural recorder is merely 5 by 5 mm [15]. As a result, to maximize the magnetic flux crossing the secondary antenna and improve the coupling factor, the size of the secondary is designed as large as possible. To the contrary, the size limitation on the primary antenna is usually not as strict as the secondary part. There is an optimized size for each primary antenna, which will be explicated later.

There are different structures to transfer wireless power. By selecting the appropriate power transfer structure, designers could optimize the antenna size and the weight. For example, the LC-pair introduced later has the smallest antenna size, when the Quad-loops introduced in this chapter require the largest space. In the same transfer structure, the higher the operating frequency is, the smaller the antenna is. However, the circuit efficiency degrades and the dielectric loss in human body increases when the frequency increases. As a result, designers have to trade off between the antenna size and other considerations.

3. Quality factor: Quality factor or Q factor is a dimensionless parameter that describes how under-damped a power antenna is. High Q indicates a low rate of energy loss relative to its stored energy in the power antenna. In this chapter, we will introduce the inductive coupling efficiency are related to the coupling strength between the quality factors of the transmitting and receiving antennas. The higher the Q factors are, the higher the transfer efficiency is.

Typically, the Q factor of on-chip coil is in the range of 5–30. The Q factor of off-chip copper coil is in the range of 10–200. To the best of our knowledge, the power antenna with the highest Q factor in wireless power transfer is the helical antenna used in strong-couplings. Its Q factor reaches 950 [8]. However, the helical antenna has much larger diameter than the regular close-loop coils. Accordingly, designers have to trade off between the antenna size and the quality factor.

For a certain coil, the Q factor varies with operating frequency. The Q factor of the coil firstly increases when the frequency increases. Then, the Q factor reaches its peak value and decreases when the operating frequency goes on increasing. When the frequency reaches the self-resonant frequency, the Q-factor decreases to zero. Accordingly, designers hope to make the coil working at the frequency that it has the maximum Q factor. It is noted that, in some applications, the Q factor is not always the higher the better. The higher the Q factor is, the narrower the bandwidth becomes. Since, the resonant frequencies of the transmitter and the receiver cannot be identically same, too narrower band may cause mistuning. The coupling efficiency consequently decreases. To solve this problem, the automatic tuning adjustment is introduced in the Chap. 5.

4. Coupling factor: Because the receiving antenna is placed over a distance from the transmitter, only a fraction of the magnetic flux generated by the transmitting coil penetrates the receiving coil. The more flux reaches the receiving coil, the better the two coils are coupled. The grade of the coupling is expressed by the coupling factor k [11].

According to the definition of the coupling factor, its value is only determined by the size and relative position of the primary and secondary power antennas. To improve the coupling factor, the primary and secondary antennas should better have the equal size and should better be placed as close as possible. However, in some biomedical microsystems, the transfer distances are much larger than the diameter of the receiving antenna. To improve the coupling, several optimization methods are to be presented in this chapter.

5. Operating frequency: The operating frequency is another extremely important design consideration in the system. The operating frequency describes the frequency of the radio waves generated by the primary coil. For sure, all power circuits and the secondary coil work at the operating frequency.

Determining the operating frequency is a comprehensive task. Designers have to consider many factors, including the antenna sizes, the Q factors, the self-resonant frequency, the circuit efficiency, and the energy absorption in human body. The antenna's size may be reduced when the frequency increases, but the circuit efficiency and energy loss get worse. The Q factor of the antenna and the overall transfer efficiency are getting better or worse can only be determined case by case. To optimize the operating frequency, designers have to make clear what the most important feature is in the system. If the overall transfer efficiency is the optimization target, an optimization method is to be presented in this chapter. Besides selecting the optimized frequency, some biomedical microsystems directly adopt a frequency in Indusial-Science-Medical (ISM) band, like 125 kHz or 13.56 MHz.

To sum up the design considerations, the key question is how to design the power antenna that could work with higher power level, have smaller size and weight, higher quality factor, higher coupling factor, and appropriate operating frequency.

3.1.3 Classification Methods

Due to the rapid development of the wireless power transfer, there have been many antenna techniques. The current power antenna techniques can be classified by several ways. In this section, we discuss the classification methods and the detailed catalog, so we could learn the power antenna from different points of view. Here, we elaborate classification methods.

(a) Classify by the coupling strength

The first way to classify the antenna techniques is the coupling strength. Existing transfer techniques can be cataloged into two types. They are the loose-couplings [16–21] and the strong-couplings [8, 22, 23].

The strong-coupling is a concept proposed in MIT's research in 2007 [8], and it is quite a hot research topic in recent years. The strong-coupling describes that wireless energy can be efficiently exchanged between strongly resonant antennas. In MIT's experiment, a transfer system with four antennas was designed. The experiment successfully transferred 60 W of power with power transfer efficiency of 40 % to a light bulb over a distance of 2 m [8]. Comparing to the strongly-couplings, the conventional inductive couplings are all called the loosely-couplings [16–21]. It is a misunderstanding that the strong-coupling has strong coupling factor k according to the literal meaning. Actually, the essential of the strong-coupling is the high Q factor. The Q factor of the power antenna adopted in the MIT's experiment is 950 [8], which is almost 10 times higher than the regular close-loop coil. According to equation introduced later in this chapter, it's not difficult to find out the high Q factor could compensate the efficiency degradation caused by the decrease of the coupling factor. In other words, the high Q factor makes the energy coupling between antennas stronger. It is noted that although the loose-couplings may also reach very high power transfer efficiency, like over 90 %, it achieves the high efficiency by the high coupling factor but not by the high Q factor. As a consequence, the efficiency of the loose-coupling increases rapidly when the transfer distance increases.

To classify the power antennas by the coupling strength is a relatively rough method. It cannot present the performance of the transfer. There are more subtypes in both the loose-coupling and the strong-coupling. Those subtypes have different features. Additionally, the loose-coupling and the strong-coupling can be mixed together to create new features for special systems. We will propose one mixed coupling transfer in this chapter. As a result, this classification method is not appropriate for this book.

(b) Classify by the number of antennas

Another way to classify the power antennas is to count the number of antennas in system. A basic power transfer system uses only two antennas. One is the primary antenna and the other is the secondary. Some systems require more power antennas, like 3–4 antennas. For example, to transfer power to farer place, more resonant antennas may be placed between the transmitting and the receiving antennas. These antennas divide the whole transfer distance into several short segments. They act like power repeaters. As another example, the strong-couplings usually adopt more than 2 antennas. In the MIT's experiment, there are totally four antennas. The first antenna is placed very near to the second antenna, and the forth antenna is near to the third antenna. The distance from the second to the third antennas is 2 m. The functions of the first and forth antennas are to deliver energy to the second and the third antennas. The second and the third antennas are the real components to transfer energy. By using the structure, the impedance matching becomes very convenient. The matching can be done by adjusting the separate distance between the first and second antennas, or the distance between the third and the forth antennas.

The problem of this classification method is it doesn't reflect the performance and the application condition of the power antennas. For the same antenna number, systems may use diverse types of antennas, like the close-loop coils or the helical antennas. Accordingly, they have entirely different performance. Therefore, in this chapter we don't go further introduce the power antennas by this way.

(c) Classify by the power transfer structures

The last way to classify the antenna techniques is the power transfer structure. The power transfer structure is a comprehensive description that describes the characteristics including the electrical features, the physical topology, and the logic relationships between the antennas adopted. According to this classification method, there are four structures. With the development of the technology, we believe there will be more structures in the future.

The classified four structures are the (1) LC-pair, (2) Multiple-resonators, (3) Quad-loops, and the (4) Helix-derivatives. The LC-pair is made up of double inductor and capacitor resonators, which is the most basic transfer structure widely adopted in the current wireless power transfer for biomedical microsystems. The structure of the Multiple-resonators is composed of at least three resonant power antennas, which is typically designed for extending the transfer distance. The Quad-loops describes the transfer structure proposed by MIT in 2007 [8], which is a representative structure for the strong-coupling using large size helical antennas. The Helix-derivatives is the last structure. It derives from the Quad-loops, but forms a new type of structure, which emphasizes the combination of high Q factor helical antennas and small size close-loop coils for biomedical applications.

The four structures are to be presented and analyzed one by one in this chapter. We choice the classification method of the transfer structure is because it is a comprehensive and concrete reflect of the features and application conditions for various power antenna systems.

3.2 LC-Pair

The LC-pair is the oldest and the most frequently adopted antenna technique in the wireless power transfer, especially for the biomedical applications. It has the smallest system size and decent power transfer efficiency over short transfer distance. Figure 3.2 shows the circuit diagram of the LC-pair. As figure shows, there are two coils. One acts as the primary coil, and the other acts as the secondary coil. Capacitor is adopted at both side to keep the coils resonate at the designed

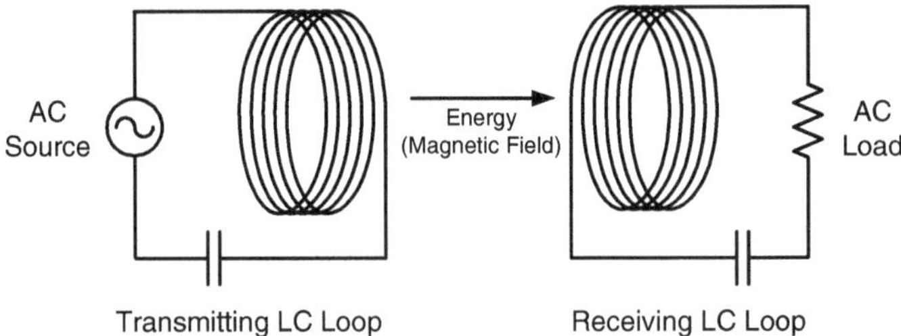

Fig. 3.2 The LC-pair in series resonance

frequency. The AC source connects to the primary coil when the AC load connects to the secondary coil.

In this section, we firstly introduce the modeling of single coil. By clearly modeling coils, the expression for the transfer efficiency is to be given. Surrounding the issue of the transfer efficiency, the optimizations of the coupling factor, quality factor, and the operating frequency are to be discussed.

3.2.1 Coil Modeling

In the LC-pair, close-loop coils are adopted as inductor to generate magnetic field in space. Before analyzing the couplings between the coils, it is necessary to firstly know the details of a single coil. Accordingly, we firstly introduce the modeling of a single coil.

Figure 3.3 shows a multiple-turn coil on the left. So far, there are many ways to model the coil [24–26]. Figure 3.3 shows a basic model. A series circuit is formed by the coil inductance *L* and a coil conductor resistor *R*. Because there are parasitic capacitors in the coil, especially the metal-to-metal capacitor between the turns, an equivalent parasitic coil capacitor C_p is connected in parallel to the inductor and the resistor. Another simplified equivalent circuit for the model is given as shown on the right in Fig. 3.3. The impedance of the coils is represented by a real impedance component *Re* and an imaginary impedance component *Im*. This modeling approach has been verified very successful, especially in the low frequency range. For those applications requires higher accuracy, like on-chip spiral coils, complex models [24–26] can be used.

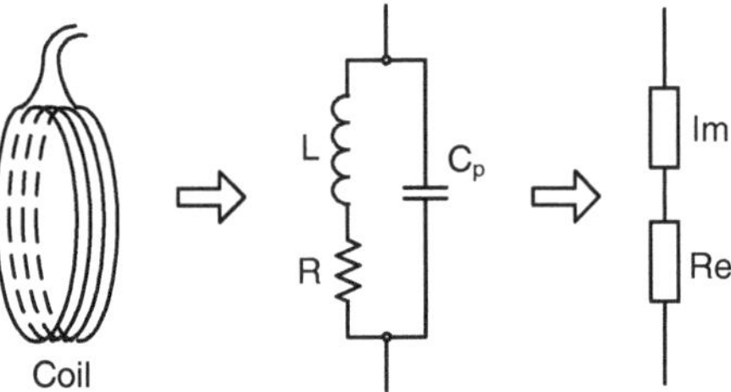

Fig. 3.3 Modeling of a single magnetic coil

(a) Coil conductor resistance

Every coil has the internal resistance R. The resistances of the transmitting and receiving coils significantly degrade the power transfer efficiency. To optimize and evaluate the transfer performance, the coil resistance must be given. We start to calculate the coil resistance by using basic physical laws. Electromagnetic simulations are also conducted to present the electromagnetic phenomenon more clearly. The equations in this section will be used later to optimize the overall performance of the wireless power transfer.

The DC resistance R_{DC} of a metal conductor is given by [11]

$$R_{DC} = \rho \frac{l}{A} \tag{3.1}$$

where l is the length of the conductor measured in meters, A is the cross-sectional area of the conductor in square meters, and ρ is the electrical resistivity of the metal material, measured in ohm-meters. This basic physical law shows the resistance inversely proportional to the cross-sectional area. Therefore, designers tend to select the wires having larger cross-sectional area. The electrical resistivity in the equation is actually a fixed constant for certain metal at certain temperature, which reflects the material's ability to oppose electric current. Here, the following shows the electrical resistivity of several common metal conductors at 20 °C.

Electrical resistivity of Silver	:	$1.59 \times 10e - 8(\Omega \cdot \mathrm{m})$
Electrical resistivity of Copper	:	$1.68 \times 10e - 8(\Omega \cdot \mathrm{m})$
Electrical resistivity of Gold	:	$1.72 \times 10e - 8(\Omega \cdot \mathrm{m})$
Electrical resistivity of Aluminum	:	$2.44 \times 10e - 8(\Omega \cdot \mathrm{m})$

Clearly, silver has the lowest electrical resistivity. Using silver, we can build coils having the lowest internal resistance at room temperature. But, it is also clear that the silver has much higher cost than copper. According to the table, the electrical resistivity of silver is 5 % smaller than the value of copper. However, silver costs approximately 34 dollars an ounce [27] when the copper costs only 0.24 dollar an ounce. As a result, most wireless power transfer systems adopt copper as the conductor for the coils.

It is noted that electrical resistivity must be given under a specified temperature. For example, the all resistivity values above are given at 20 °C. During the power transfer, the temperature may rises because the electric energy converts to heat due to the existence of the conductor resistance. Therefore, to exactly evaluate the coil resistance in the wireless power transfer, designers need considering the temperature using the following equation [11].

$$R_{DC}(T) = R_{DC}(T_0) \cdot (1 + \alpha \cdot \Delta T) \tag{3.2}$$

where α is the temperature coefficient of resistance per degree of temperature.

Besides DC resistance, there is AC resistance in the coil. The skin effect [11] increases the conductor resistance during the increase of the operating frequency.

Skin effect is the tendency of an alternating electric current to become distributed with a conductor such that the current density is largest near the surface of the conductor [11]. Because the electric current flows mainly at the skin of the conductor, the equivalent cross section of the conductor decreases. As a result, the skin effect causes the increase of the resistance at high frequency. The skin depth is defined as the depth beneath the surface of a conductor to where the current density drops to one neper between the current density at the surface. The skin depth can be approximated as the following equation [11]

$$\delta = \sqrt{\frac{2 \cdot \rho}{\omega \cdot \mu}} \tag{3.3}$$

where ρ is the resistivity of the conductor, ω is the angular frequency of the alternating current, and μ is the absolute magnetic permeability. As you can see, the skin depth varies as the square root of the resistivity of the conductor. In other words, the skin depth is thinner in a good conductor. However, it is noted that a good conductor still shows a smaller resistance. To show the skin effect clearly, we conducted a simulation as Fig. 3.4 shows.

In Fig. 3.4, there is a conductor with alternating current flowing. The grey degree shows the current density in the cross section of the conductor is not uniform. The alternating current is more like to flow near the surface of the conductor. The peripheral arrows show the transient magnetic field generated by the alternating current. Due to the skin effect, the AC resistance R_{AC} of a wire in length l is donated as [11]:

$$R_{AC} = \frac{l \cdot \rho}{\pi \cdot D \cdot \delta} \tag{3.4}$$

where D is the diameter of the wire, and δ is the skin depth. This equation assumes $D \gg \delta$.

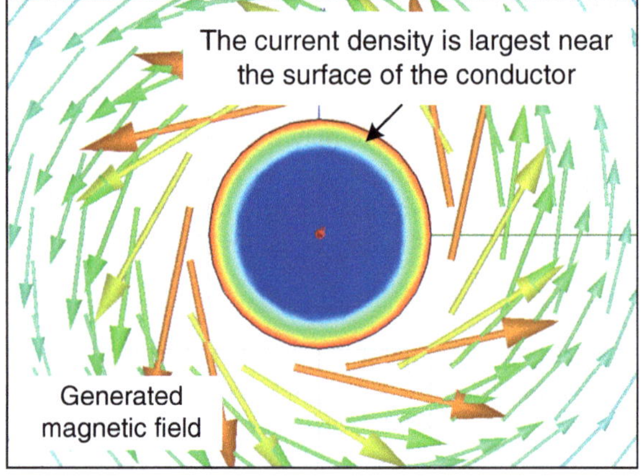

Fig. 3.4 The current distribution in a wire cross section and the generated magnetic field

Another electromagnetic effect that may be considered for the AC resistance is the proximity effect [11]. If current flows through one or more other nearby conductors, the distribution of the current in the first conductor will be constrained to smaller regions. In the wireless power transfer, the current usually flows in multiple turn coils. Consequently, the current tends to concentrated in the areas of the center of the wire. This effect conflicts with the skin effect. Accordingly, the two effects make the equivalent sectional area of the high frequency current much smaller than the area of the low frequency current. We conducted a simulation as Fig. 3.5 shows.

In Fig. 3.5, there are two wires in parallel. Two AC currents flow in the wires with the same current direction. According to Fig. 3.5, the current density at where nearby the other wires is relatively low. This is the proximity effect. Since the expression of the proximity effect is quite complex, the AC resistance usually only counts the skin effect. To reduce the effects, the Litz wire can be adopted. It's a type of cable consisting of many individually insulated thin wire strands.

A DC resistance equation and an AC resistance equation have been given respectively. Which equation is going to be adopted for the coil model is determined by the relationship between the radius of the wire strip and the calculated skin depth. When the calculated skin depth is smaller than the radius of the wire strip, the AC resistance equation should be employed instead of the DC resistance equation.

(b) Coil inductance

To model the coil precisely, the coil inductance and the mutual inductance are be evaluated. Here, we cited the equations from the radio engineer's handbook [11]. All inductances are given for low frequency condition.

The self-inductance L of a single-turn circular loop is given by [11]

$$L = 4.86 \times 10^{-3} \cdot D \cdot \left(2.303 \log_{10} \frac{8 \cdot D}{d} - 2 + \mu \cdot \delta\right) \tag{3.5}$$

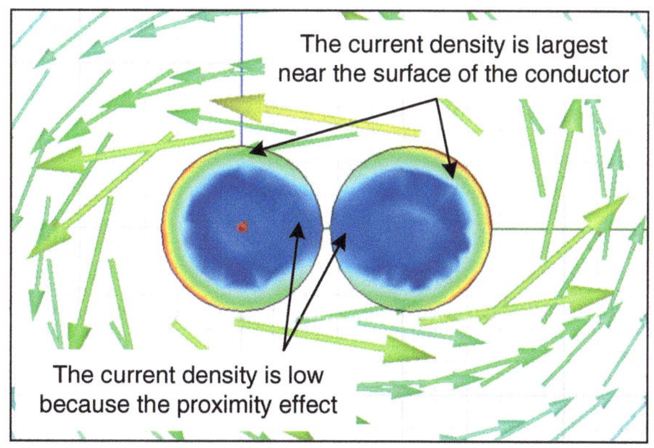

Fig. 3.5 The current distribution in two parallel wire cross sections and the magnetic field

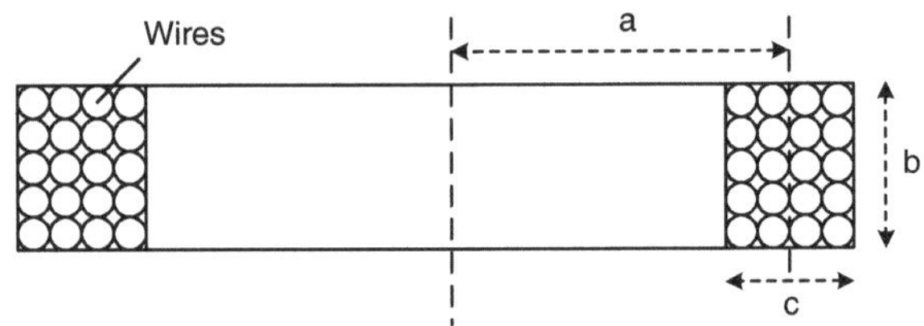

Fig. 3.6 Circular multiple-layer multiple-turn coil

where D is the diameter of the circular loop, d is the wire diameter in meters, μ is the absolute magnetic permeability of the conductor, and δ is a skin effect correction factor as a function of wire diameter and frequency.

The self-inductance L of a single-layer coil is given by [11]

$$L = F \cdot n^2 \cdot d \tag{3.6}$$

where d is the diameter of the coil in inches, n is the number of turns, and F is a quantity that depends upon the ratio of the diameter to the length of the coil. The determination of the F uses table look-up method.

Most of the wireless power transfer systems use circular multiple-layer coils as Fig. 3.6 shows. The parameter a is the average radius, b is the axial length, and the c is the radial thickness of the winding.

The self-inductance L of the multiple-layer coil is given by [11]

$$L = F \cdot n^2 \cdot d - \frac{0.013193 \cdot n^2 \cdot a \cdot c}{b} \cdot (0.693 + B_s) \tag{3.7}$$

where F is a quantity that depends upon the ratio of the diameter to the length of the coil, which can be find by using table lookup method, the n is the number of turns, B_s is a function of wire length and the radial thickness c of wire. B_s also uses table lookup method to get its value.

A simple approximate equation for the multiple-layer inductance of the coils is given by [11]

$$L = \frac{0.8 \cdot a^2 \cdot n^2}{6a + 9b + 10c} \tag{3.8}$$

where n is the number of turns, a, b, and c are the physical dimension of the coil as mentioned above. All dimensions are in inches.

(c) Coil parasitic capacitor

Every coil has parasitic capacitor. We analyze the parasitic capacitor because it has significant effect on the frequency characteristic of the coil.

A simplified equation for the parasitic capacitor is given by [11]

$$C_{Self} = \sum_{p<k} C_{p,k}(k-p)^2 \Big/ N_t^2 \tag{3.9}$$

where $C_{p,k}$ is the parasitic capacitance between the coil winding layer p and the layer k. The parasitic capacitance inside of a layer is ignored because the voltage difference inside of the same layer is relative much smaller.

Due to the existence of the parasitic capacitor, the high frequency alternating current in the coil is changed by the capacitor. When the current frequency is high enough, most current in the coil would be bypassed via the parasitic capacitor. As a result, the coil cannot be viewed as a lumped inductor anymore. The coil's inductance would resonate with the parasitic capacitor at a frequency. That frequency is called the self-resonant frequency. It is given by:

$$f_{Self} = \frac{1}{2\pi\sqrt{LC_{Self}}} \tag{3.10}$$

(d) Coil impedance in real and imaginary components

As Fig. 3.3 shows, the impedance of a coil can be donated as a serial connection of a real and an imaginary component. The real component represents all energy loss in the coil, and the imaginary component represents all energy storage in the coil. The impedance of the coil can be given by

$$Z = \frac{R + j\omega L\left(1 - \omega^2 LC - \frac{CR^2}{L}\right)}{\left(1 - \omega^2 LC\right)^2 + \omega^2 C^2 R^2} = Z_{\mathrm{Re}} + jZ_{\mathrm{Im}} \tag{3.11}$$

where R, L and C is the coil resistance, inductance, and parasitic capacitance respectively. The real component of the coil impedance Z_{Re} is given by

$$Z_{\mathrm{Re}} = \frac{R}{\left(1 - \omega^2 LC\right)^2 + \omega^2 C^2 R^2} \tag{3.12}$$

And the imaginary component of the coil impedance Z_{Im} is donated as

$$Z_{\mathrm{Im}} = \frac{\omega L\left(1 - \omega^2 LC - \frac{CR^2}{L}\right)}{\left(1 - \omega^2 LC\right)^2 + \omega^2 C^2 R^2} \tag{3.13}$$

The quality factor or Q factor for a coil is a dimensionless parameter that describes how under-damped a coil is. It is defined as the ratio of imaginary component to the real component. It's given as:

$$Q = \frac{|Z_{\mathrm{Im}}|}{Z_{\mathrm{Re}}} = \frac{\left|\omega L\left(1 - \omega^2 LC - \frac{CR^2}{L}\right)\right|}{R} \tag{3.14}$$

So far, we have presented all circuit parameters for a single coil. Most of these parameters vary with the operating frequency. Accordingly to better understand the behavior of a coil with regard to the operating frequency, four figures are given in Fig. 3.7. They are the conducting resistance, real impedance, imaginary component, and the quality factor.

As Fig. 3.7 shows, the conducting resistance continuously increases with the increase of the frequency. The real impedance reaches its maximum value and the imaginary impedance reaches zero when the coil works at the self-resonant frequency. The result of the quality factor shows that it firstly increases and then decreases. The quality factor decreases to zero when the operating frequency reaches the self-resonant frequency. Clearly, there is an optimized frequency for the quality factor to reach the peak value. Designed coils are usually supposed to work at that frequency.

3.2.2 Expression of the Efficiency

The power transfer efficiency of the LC pair can be calculated by several ways [11, 19, 20]. Here, we present a classic way [28], which calculates the efficiency by

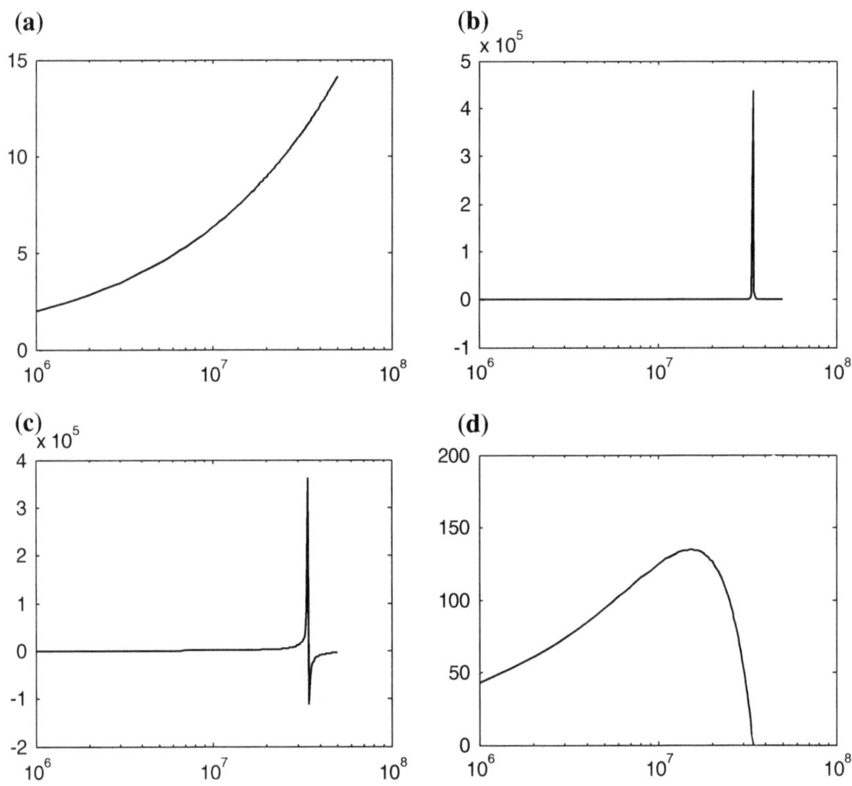

Fig. 3.7 The frequency response of a coil

reflecting the load resistance R_L at the secondary side to an equivalent resistance R_e at the primary side, which is given by

$$R_e = \frac{(\omega M)^2}{R_2 + R_L} \tag{3.15}$$

where M is the mutual inductance between the primary and the secondary coils. R_2 is the internal resistance of the secondary coil. R_L is the resistance of the load. Suppose there is AC power source at the primary side, and it outputs the power of $P_{emitted}$. As a result, the power transferred to the load $P_{transferred}$ can be calculated as:

$$P_{transferred} = \frac{R_L}{R_2 + R_L} \cdot \frac{R_e}{R_1 + R_e} \cdot P_{emitted} \tag{3.16}$$

where R_1 is the internal resistance of the primary coil. In this equation, it is noted that we have assumed that both the primary and the secondary coils have been tuned with capacitors for a designed operating frequency. According to the definition of the quality factor and the coupling factor, the internal resistances of the primary coil and the secondary coil and the mutual inductance are given by:

$$R_1 = \frac{\omega L_1}{Q_1}, \ R_2 = \frac{\omega L_2}{Q_2}, \ M = k\sqrt{L_1 L_2} \tag{3.17}$$

Substituting Eq. (3.17) into (3.16), the transfer efficiency can be donated by the coupling factors and quality factors instead of the internal resistance and the mutual inductance. According to previous research [28], the overall power transfer efficiency can be optimized when the internal resistance of the secondary coil equals:

$$R_2 = R\frac{\sqrt{1 + k^2 Q_1 Q_2}}{Q_2^2} \tag{3.18}$$

As a result, substituting Eq. (3.18) into (3.16), the efficiency η reaches the maximum value:

$$\eta = \frac{P_{transferred}}{P_{emitted}} = \frac{k^2 Q_1 Q_2}{(1 + \sqrt{1 + k^2 Q_1 Q_2})^2} \tag{3.19}$$

Clearly, the optimized transfer efficiency is only determined by the factor $k^2 Q_1 Q_2$. Therefore, designers try their best to increase the coil's k and Q factors. Due to coupling factor k is only determined by the shape, size and the relative position of the coils, designers usually firstly optimize the coupling factor, and then optimize the Q factor.

3.2.3 Optimization of the Coupling Factor

According to the previous modeling, we have found two issues of crucial importance: the coupling factor and the quality factor. And we have known designers usually optimize the coupling factor before other parameters. This section introduces the optimization of the coupling factor.

In most of biomedical applications, there is usually a very strict size limitation on the secondary side. For example, the size of the capsule endoscopy is approximately 11 by 27 mm. To increase the magnetic flux penetrating the secondary coil, the secondary coil is designed as large as possible within the limit. Meanwhile, the transfer distance is usually specified by applications. For example, the transfer distance for the capsule endoscopy is in range from 1 to 30 cm to cover the whole abdomen area. As a result, there is only one option left for the designers. It's the diameter of the primary coil. Therefore, to increase the coupling factor, designers turn to optimize the size of the primary coil. Figure 3.8 shows a circular primary coil and the generated magnetic field strength H over a distance of x.

Aiming the above figure, we present 4 methods to optimize the coupling factor between the primary coil and an assumed small size secondary coil over a transfer distance of x. The first way optimizes the coupling factor by calculating the magnetic field strength H. The second way to optimize the coupling is calculating the mutual inductance. The third way takes account the consideration of coil resistance. As a result, the result is slightly different from the two previous methods. The last method is presented for those applications having moving receivers.

1. Optimization method 1

The first optimization method has a very straight train of thought. The more magnetic flux generated by the primary coil penetrates the secondary coil, the higher the coupling factor is. Accordingly, we try to maximize the magnetic field strength over the distance of x. The magnetic field strength at distance x is given by:

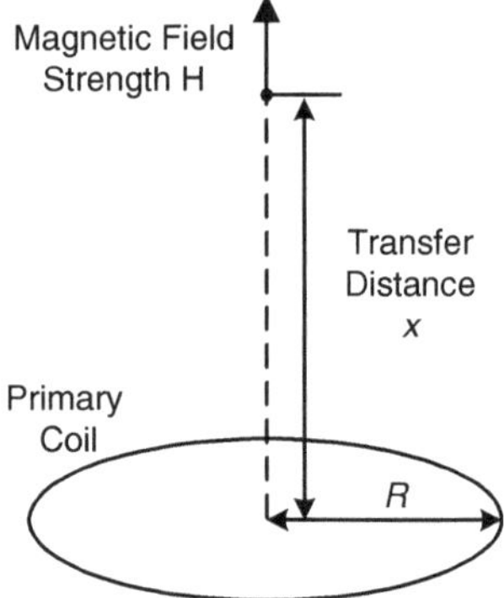

Fig. 3.8 A primary coil and the generated magnetic field

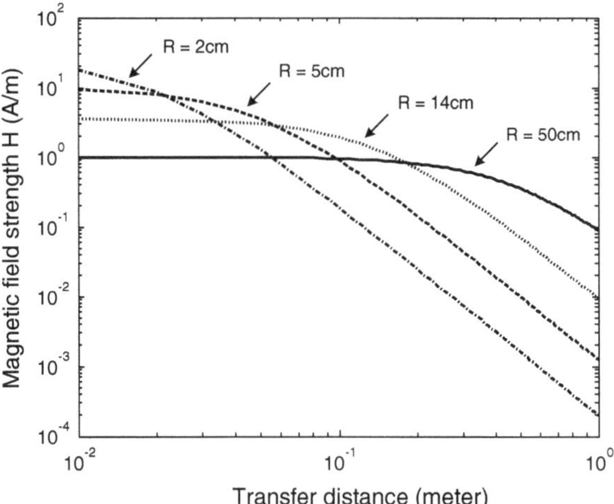

Fig. 3.9 Magnetic field strength generated by primary coils in different radiuses

$$H = \frac{I \cdot N \cdot R^2}{2\sqrt{(R^2 + x^2)^3}} \tag{3.20}$$

where I is the current magnitude in the primary coil, N is the number of turns, R is the radius of the primary coil, and x is the transfer distance. Figure 3.9 shows the calculated magnetic field strength with regards to the transfer distance and the radius of the primary coil.

In the figure, there are several lines. They are calculated using different radius of the primary coil. Their corresponding radiuses R are 2, 5, 14, and 50 cm respectively. Each line is the calculated magnetic field strength H with regards to the transfer distance.

According to the figure, it is clear that when the radius of the primary coil is relatively small, the magnetic field strength at near place is very strong. To the contrary, when the radius of the primary coil is relatively large, the magnetic field strength at near place is weaker but the magnetic field strength is stronger at far place. Since we would like the magnetic field strength as strong as possible over a specified transfer distance, we are going to find out the optimized radius of the primary coil for a specified transfer distance. For example, if we would like to transfer the power over a distance of 0.1 m. According to the figure, when the radius of the primary coils is about 14 cm, the magnetic field strength has the highest value over a transfer distance of 0.1 m.

To clearly present the different performance at different radius, Fig. 3.10 shows another group of lines. There are four lines in the figure. Their transfer distances are 10, 20, 30, and 40 cm respectively. Each line represents the change of the magnetic field strength with regards to the radius of the primary coil. According to the figure, there are clearly two laws. One is the value of the peak efficiency decreases when the transfer distance increases. The other is the corresponding optimized radius of the primary coil increases when the transfer distance increases.

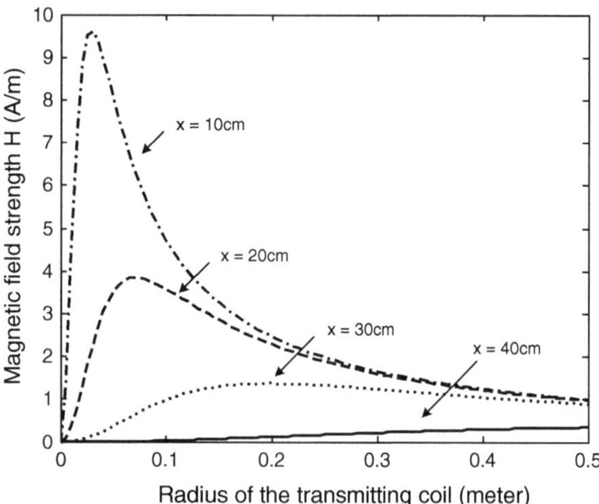

Fig. 3.10 Magnetic field strength generated by primary coils at different distance

To analyze this phenomenon precisely, the following equation is given:

$$\frac{dH(R)}{dR} = \frac{2 \cdot I \cdot N \cdot R}{\sqrt{(R^2 + x^2)^3}} - \frac{3 \cdot I \cdot N \cdot R^3}{(R^2 + x^2) \cdot \sqrt{(R^2 + x^2)^3}} \tag{3.21}$$

To maximize the magnetic field strength H, the radius R of the primary coil needs to be 1.414 times of the transfer distance,

$$R_{1(Optimal)} = \sqrt{2} \cdot x \tag{3.22}$$

The maximal value of the magnetic field strength H is given by:

$$H_{Optimal} = \frac{I \cdot N}{\sqrt{3} \cdot x} \tag{3.23}$$

As Eq. (3.20) shows, the magnetic field strength H originally follows an inverse-cube law with regards to transfer distance for a radius fixed primary coil. After the optimization, the magnetic field strength H has been optimized to be inversely proportional to the transfer distance as Eq. (3.23) shows. This is the first optimization method.

2. Optimization method 2

The second way is to optimize the mutual inductance. To calculate the mutual inductance, the magnetic flux B is to be expressed firstly. Because the radius of the receiving antenna is usually much smaller than the separate distance, we assume that the magnetic flux density B penetrating the receiving antenna is uniform. Thus, the magnetic flux Φ_{21} can be calculated by a product of the magnetic flux density B and the area of the receiving antenna.

$$\Phi_{21} = \iint |B| \cdot dS = \frac{\mu N_1 I_1}{2} \cdot \frac{R_1^2}{(R_1^2 + x^2)^{3/2}} \cdot \pi R_2^2 \tag{3.24}$$

where I_1 is the current magnitude in the primary coil, N_1 is the number of turn of the primary coil, R_1 and R_2 is the radius of the primary and secondary coils respectively, x is the transfer distance, and μ is the permeability. The mutual inductance is given by:

$$M = \frac{N_2 \Phi_{21}}{I_1} = \frac{\mu \cdot N_1 N_2}{2} \cdot \frac{R_1^2}{(R_1^2 + x^2)^{3/2}} \cdot \pi R_2^2 \tag{3.25}$$

where N_2 is the number of turn of the secondary coil. In order to find the optimized radius of the primary coil for the mutual inductance, the following differential equation is given:

$$\frac{\partial M}{\partial R_1} = \pi R_2^2 \cdot \frac{\mu \cdot N_1 N_2}{2} \cdot \frac{2R_1 \left(R_1^2 + x^2\right)^{3/2} - 3R_1^3 \left(R_1^2 + x^2\right)^{1/2}}{\left(R_1^2 + x^2\right)^3} \tag{3.26}$$

By making the above equation equal to zero, the optimal radius of the primary coil is found:

$$R_{1(Optimal)} = \sqrt{2} \cdot x \tag{3.27}$$

This optimization method has the same conclusion as the first optimization method introduced above. It's because their essential target is the same. It's the ratio of the magnetic flux penetrating the secondary coil to the magnetic flux generated by the primary coil.

3. Optimization method 3

In the first two methods, we didn't consider the relationship between the coupling factor and the quality factor. As introduced in the expression of the efficiency, the ultimate target we need to optimize is the following factor:

$$k^2 Q_1 Q_2 \tag{3.28}$$

The coupling factor k and the quality factor Q in the above equation are actually not independent. For example, when we change the radius of the primary coil for maximizing the coupling factor, its quality factor is also altered. Accordingly, we present the third optimization method, in which we determine the radius of the primary coil by maximizing the whole factor of $k^2 Q_1 Q_2$. First, we express the factor by the mutual inductance and other parameters as follows:

$$k^2 Q_1 Q_2 = \frac{(\omega M)^2}{Z_1 Z_2} \propto \frac{M^2}{R_1 R_2 N_1 N_2} \tag{3.29}$$

where Z_1 and Z_2 are the resistances, R_1 and R_2 are the radiuses, and N_1 and N_2 are the turns of the primary and secondary coils respectively. Substituting Eq. (3.25) into (3.29), a new expression is given as:

$$k^2 Q_1 Q_2 \propto \frac{1}{4}\mu_0^2 \pi^2 N_1 N_2 \cdot \left(\frac{R_1 R_2}{R_1^2 + x^2}\right)^3 \tag{3.30}$$

To optimize the radius R_1 of primary coil, the differential equation is given:

$$\frac{\partial\left(\frac{R_1 R_2}{R_1^2 + x^2}\right)}{\partial R_1} = R_2 \cdot \frac{x^2 - R_1^2}{\left(x^2 + R_1^2\right)^2} \tag{3.31}$$

According to the above equation, if r_1 equals to D,

$$R_{1(Optimal)} = x \tag{3.32}$$

the efficiency has the maximal value. It is noted that this result is different to the first two methods. It's because this method takes account into the consideration of the quality factor.

4. Optimization method 4

All optimization methods above are given for the wireless power transfer that a fixed transfer distance is specified. However, in some applications, the transfer distance is changeable. For example, because the endoscopic capsule moves in abdomen area of human body, the transfer distance between the primary coil and the capsule is changeable, typically in range from 1 to 30 cm. The coupling factor should be optimized with consideration for the whole transfer range but no only for a fixed point.

In this optimization method, we try to find the best primary coil radius for the whole transfer range. Our method is to maximize the integration of the magnetic field strength in the whole specified transfer range. The integration of the magnetic field strength can be given by:

$$\int_{x1}^{x2} H dx = \int_{x1}^{x2} \frac{I \cdot N \cdot R^2}{2\sqrt{\left(R^2 + x^2\right)^3}} dx \tag{3.33}$$

where I is the current in the primary coil, R is the radius of the primary coil, x_1 and x_2 are the start and the end points of the transfer range. By using this equation, we get a group of lines as Fig. 3.11 shows.

The lines in the above figure are the calculated integration of the magnetic field strength in specified transfer ranges. They are calculated and plotted by Matlab. For example, the top line integrates the magnetic field strength in the transfer distance range from 0 to 15 cm, and the second top line integrates from 1 to 15 cm. For each line, it firstly increases and then decrease. Accordingly, there will

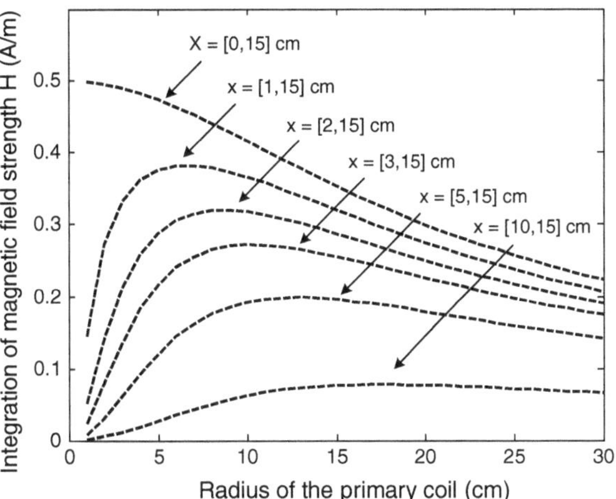

Fig. 3.11 Integration of the magnetic field strength in a specified transfer range

be an optimized radius of the primary coil for each transfer range. Using figures like this, we could get a sum of the field strength in a transfer range and find the optimized primary coil radius. For instance, the optimized radius for the transfer range from 5 to 15 cm is approximately 12 cm. Besides the integrated strength, an average strength for a specified range can be calculated. To sum up, this method could help designers finding out the best radius of the primary coil for a specified transfer range.

3.2.4 Optimization of the Q Factor and the Frequency

So far, we have introduced the optimization of the coupling factor. As introduced in the expression of the efficiency. Another optimization target is the quality factor.

The Q factor follows a complex function with regards to the operating frequency. Although it is very difficult to perform the theoretical optimization of the Q factor, we present a group of simulated and measured results in this section for its optimization. Figure 3.12 shows the simulated Q factors of a group of coils. The coils are assumed having the same diameter but different turns.

The key question for the above figure should be how to find the best number of coil turns and the best operating frequency so the Q factor has the highest value. As shown in Fig. 3.12, if the coil has only a few turns, the coil inductance would be relatively small to restore enough magnetic energy in the coil. As a result, the Q factor is not satisfied. The benefit is the parasitic capacitor would be also small. So the coil is able to work in a broad frequency band. To the contrary, if the coil has too many turns, the resistance and the parasitic capacitor would be too larger. The coil has to work in a low frequency range and the Q factor is also not satisfied. Since the radius of the primary coil has been determined by the previous

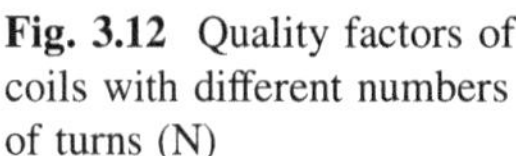

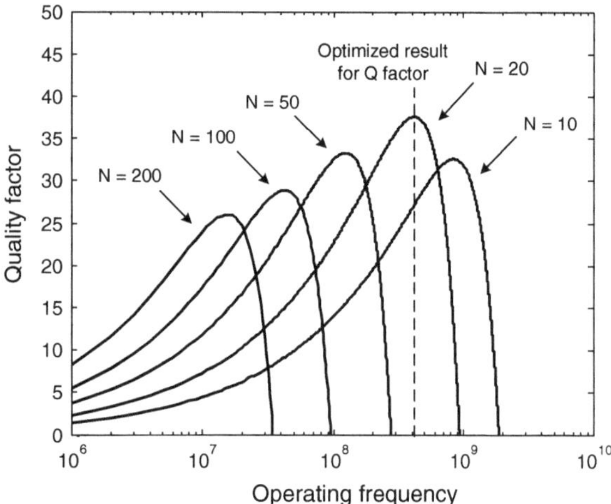

Fig. 3.12 Quality factors of coils with different numbers of turns (N)

optimization, the best way to find the highest Q factor is experiment. The following table shows a group of experimental results measured by using a network analyzer.

As Table 3.1 shows, the measured parameters include frequency, resistance, inductance, and Q factor. For this coil, the highest Q factor is 132 when the operating frequency is 20 MHz. To find out the best turns, more experiments were conducted on a group of coils. They have the same diameter but different turns.

As the Table 3.2 shows, we conducted eight experiments. The purpose is to find out in which turns, the coil has the highest Q factor, and what the operating frequency is. In the table, it's clear that when the coil has more turns, its operating frequency range reduces. The single-turn coil is able to work at up to 20 MHz, but the 8-turns coil can work at most 1 MHz. It's because the multiple turns coils have larger parasitic capacitor and lower self-resonant frequency. According to the

Table 3.1 A group of experimental results showing the relationship of the operating frequency and the Q factor

Circuit parameters and the Q factor for a single turn coil (N = 1)				
Frequency (Hz)	R (Ohm)	X (Ohm)	L (μH)	Q
100 K	0.46	0.40	0.64	0.9
500 K	0.20	1.2	0.37	5.9
1 M	0.39	2.6	0.42	6.7
2 M	0.31	5.2	0.41	17
5 M	0.48	13	0.42	28
10 M	0.44	27	0.43	60
15 M	0.59	44	0.46	74
20 M	0.50	66	0.53	132

Note All results are measured on a coil in diameter of 1 cm by using a network analyzer. The used copper wire is in the diameter of 0.5 mm

Table 3.2 A group of experimental results showing the relationship of the number of turns and the Q factor

Quality factors for coils with different turns N								
Frequency (Hz)	N=1	N=2	N=3	N=4	N=5	N=6	N=7	N=8
100 K	0.9	2.4	4.2	3.5	9.4	6.3	16	24
500 K	5.9	36	40	21	62	30	62	70
1 M	6.7	22	34	28	60	46	93	100
2 M	17	91	49	49	78			
5 M	28	61	88	114		Coils change from inductive to capacitive		
10 M	60	132	106	49				
15 M	74	139	65					
20 M	132							

Note All results are measured on a coil in diameter of 1 cm by using a network analyzer. The used copper wire is in the diameter of 0.5 mm

table, the highest Q factor is 139 when the number of turns is 2 at the operating frequency of 15 MHz.

Although we have found the highest Q factor for the coil and the corresponding operating frequency, they're not the final optimization result. It's because we haven't considered the efficiency degradation caused by the electromagnetic absorption in human body and the power conversion efficiency of circuits in system.

Figure 3.13 shows the degradations. Clearly, the higher the operating frequency is, the more the degradation is. It's because both the energy loss in the human tissue and in the circuits increases when the frequency increases.

How to estimate the degradations? Actually, designers could find conclusions in previous researches as the references. For example, when the transmitting power is 0.9 W, the localized SAR is around 0.3 W/kg at 400 kHz and 4.5 W/kg at 700 kHz [29]. To consider the degradation or not would cause entirely different optimization results. Additionally, designer should count into the degradation caused by the power conversion efficiency of the circuits. For example, rectifiers working at the frequency of 900 MHz to 1 GHz usually have power efficiency in range from 36 to 65 % [4, 5]. For those rectifiers working at a relatively low frequency, like 13.56 MHz, the efficiency is usually in the range from 45 to 80 % [2, 3].

When using the degradation method, it's noted that designers could firstly multiplies the Q factor of the primary coil by the according Q factor of the secondary coil. Then the process of the degradations can be conducted. By doing so, a balance between the optimal frequencies of the primary and secondary sides can be achieved. Usually, the primary side has a lower optimized frequency than the secondary side because the primary coil has larger size.

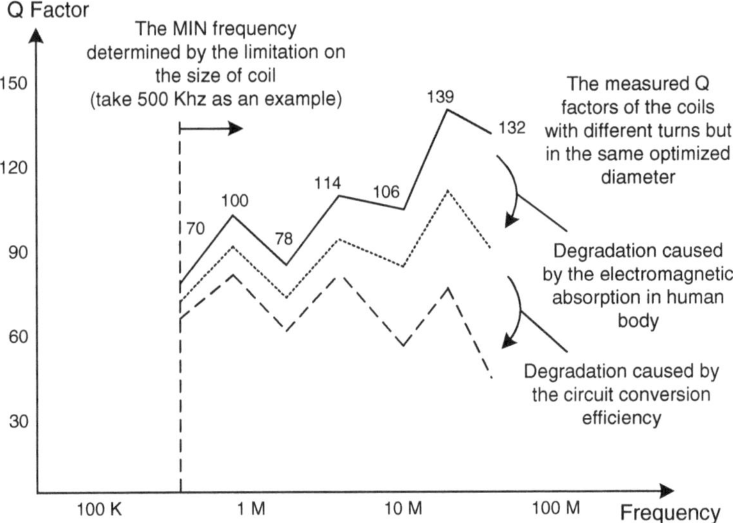

Fig. 3.13 Degradation of the quality factors caused by the electromagnetic absorption in human body and the circuit conversion efficiency

To sum up the whole optimization methods above, we have introduced the following steps:

1. The first step is to determine of the maximal radius of the secondary coil according to the size limitation on the secondary side. Usually, the secondary coil should be designed as large as possible.
2. The second step is to optimize the radius of the primary coil. We have introduced 4 methods according to different application conditions. Their essential target is to optimize the coupling factor between the primary and the secondary coils.
3. The third step is to conduct a group of experiment. Different turns of coils are needed to be measured for recording their Q factors in frequencies. These coils have different turns but in the same radius as optimized in Step 2. Both the primary and secondary coils are needed to do so.
4. The forth step multiplies the Q factors of the primary coil by the Q factors of the secondary coil, which makes sure both of the primary and secondary sides are considered.
5. The fifth step degrades the product of the Q factors by considering the electromagnetic absorption in human tissue and the circuit conversion efficiency. The percentage of the degradation varies with frequency and power level in system. Designers could find conclusions in previous researches as references.
6. The last step finds out the max degraded product of the Q factors. The according turns of the primary and secondary coil are the finally optimized results. The corresponding frequency is the best operating frequency for

system, which has comprehensively considered the system size, the coupling efficiency, the electromagnetic absorption, and the circuit conversion efficiency.

To conclude this section, we have introduced the wireless power transfer structure of the LC-pair. The coil modeling, the expression of the efficiency, the optimizations of the coupling factor, the quality factor, and the operating frequency are all introduced. The most of the optimization methods in this section can be applied to other wireless power transfer structures introduced later. It's because the LC-pair is the most basic transfer structure and the magnetic loop coil is the most common power antenna in the wireless power transfer. In following sections, basic parameters and basic optimization methods would not be repeated. We are to emphasize the differences between the structures.

3.3 Multiple-Resonators

As introduced in Sect. 3.1, this chapter introduces the power antennas by the classification of the transfer structures. In this section, we introduce the transfer structure of the Multiple-resonators.

3.3.1 Introduction

Unlike the LC-pair, the Multiple-resonator is a transfer structure that adopts at least three electromagnetic resonators to transfer wireless power from a source to a load. Figure 3.14 shows the transfer structure.

As shown in Fig. 3.14, the source connects to a primary resonator and the load connects to a secondary resonator. All resonators in the middle are named middle resonators. The resonator can be any types of components that connect electrical energy in circuit to magnetic energy in space. These resonators are all tuned at the same operating frequency, so the impedances are matched. Our previous research [30] focused on this topic.

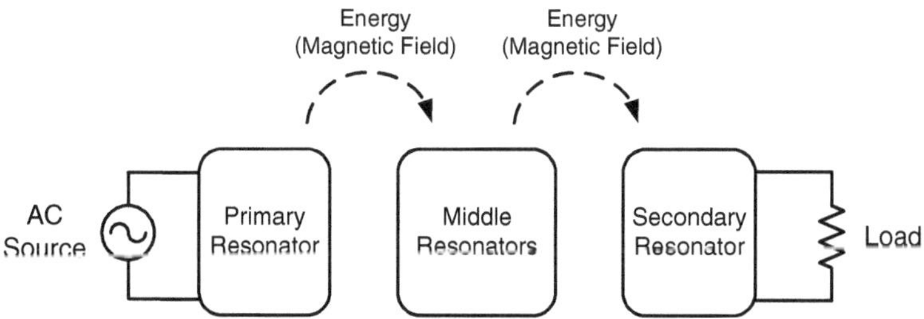

Fig. 3.14 The Multiple-resonators

There are two advantages of the Multiple-resonator. (1) The first is extended transfer distance. As introduced, the coupling factor in LC-pair decreases when the transfer distance increases. As a consequence, the transfer efficiency degrades. If one resonator can be inserted between the primary and secondary sides, the original small coupling factor can be replaced by a product of two much larger coupling factors. The first coupling factor k_{12} present the coupling between the primary resonator and the middle resonator. And the second coupling factor k_{23} describes the coupling between the middle resonator and the secondary resonator. Although the internal resistance of the middle resonator consumes energy, it may contribute more to the enhancement of the couplings. (2) The second advantage is the middle power may have a relatively higher Q factor than the primary and the secondary sides by using heavy wires or other techniques. According to previous Eq. (3.19), the high Q factor can be used to compensate the week coupling factor and promote the power efficiency. The main disadvantage of the Multiple-resonator is that the increased middle resonator occupies space. Since the most of biomedical microsystems have limitation on the size of the secondary side, the middle power resonator is usually placed outside of human body.

3.3.2 Detailed Designs

In this section, we introduce a detailed design of the Multiple-resonator. This design adopts three LC circuits as the electromagnetic resonators. Both theoretical analysis and experiments are to be presented. The detailed structure is shown in Fig. 3.15.

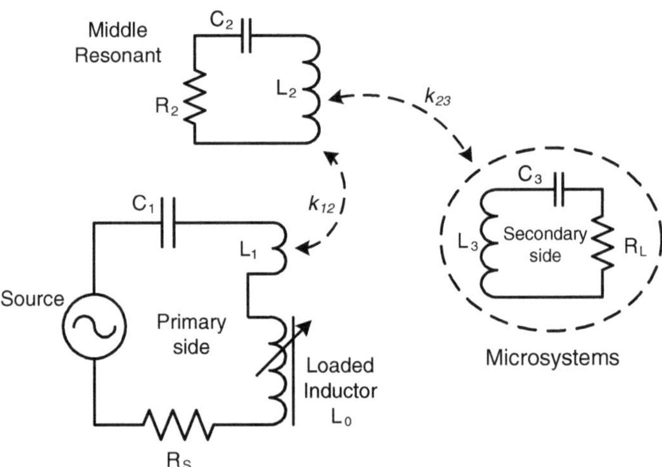

Fig. 3.15 A detailed design of the Multiple-resonators

According to Fig. 3.15, there are totally three LC resonators in the transfer structure. The three coils are nominated the primary coil, middle resonant coil, and secondary coil respectively. At the primary side, there are the five components. They are the AC power source, primary coil L_1, loaded inductor L_0, tuning capacitor C_1, and equivalent internal resistance R_s for the whole primary circuit. There are two inductors in the primary side because the primary coil L_1 usually needs only a few turns to couple energy to the middle power resonator. As a result, a loaded inductor L_0 is required to make sure the circuit is tuned at the designed operating frequency. By using a ferrite core inductor as the loaded inductor, we can easily adjust its inductance for the matching the impedance.

In the middle power resonator, there are three components. They are the inductance L_2, the tuning capacitor C_2, and the equivalent internal resistance R_2. Due to the middle power resonator typically uses heavy wires, the internal resistance R_1 is very small, like less than 1 Ohm. Accordingly, the middle resonator is expected to be a high Q factor circuit. The most energy transmitted from the primary side would be stored in the magnetic field nearby the middle resonator, and would dissipate little on its resistor.

At the secondary side, there are three components including inductance L_3, tuning capacitor C_3, and internal resistance R_L. The whole transfer process can be viewed as two steps. The first step transfers energy from the primary coil to the middle resonator by coupling factor k_{12}. The middle resonator acts like an "energy pool" in space. The secondary side draws energy from the middle resonator to the load. The coupling directly from the primary to the secondary sides are neglected to simplify the analysis.

According to circuit theory, the current I_{Load} at the secondary side is given by:

$$I_{Load} = \frac{U_s \cdot \omega^2 M_{12} M_{23}}{(R_S + R_1)R_2(R_3 + R_L) + \omega^2 M_{12}^2 (R_3 + R_L) + \omega^2 M_{23}^2 (R_S + R_1)} \tag{3.34}$$

where U_S is the voltage of the AC power source, ω is the angle frequency, M_{12} and is the mutual inductance between the primary coil and the middle resonator. M_{23} is the mutual inductance between the middle resonator and the secondary coil. According to the definition of the Q factors and the coupling factors, the following equations are given.

$$R_1 = \frac{\omega L_1}{Q_1},\ R_2 = \frac{\omega L_2}{Q_2},\ R_3 = \frac{\omega L_3}{Q_3} \tag{3.35}$$

$$M_{12} = k\sqrt{L_1 L_2},\ M_{23} = k\sqrt{L_2 L_3} \tag{3.36}$$

Substituting Eqs. (3.35) and (3.36) into (3.34), the current on the load I_{Load} can be given by:

$$I_{Load} = \frac{U_s \ \ \omega^2 k_{12} k_{23} L_2 \sqrt{L_1 L_3} \cdot Q_1 Q_2 Q_3}{\omega^3 L_1 L_2 L_3 + \omega^3 k_{12}^2 Q_1 Q_2 L_1 L_2 L_3 + \omega^3 k_{23}^2 Q_2 Q_3 L_1 L_2 L_3} \tag{3.37}$$

$$= \frac{k_{12}k_{23} \cdot Q_2\sqrt{Q_1Q_3}}{1 + k_{12}^2Q_1Q_2 + k_{23}^2Q_2Q_3} \cdot \frac{U_s}{\sqrt{(R_S + R_1)(R_3 + R_L)}} \tag{3.38}$$

$$\leq \frac{k_{12}k_{23} \cdot Q_2\sqrt{Q_1Q_3}}{1 + 2k_{12}k_{23} \cdot Q_2\sqrt{Q_1Q_3}} \cdot \frac{U_s}{\sqrt{(R_S + R_1)(R_3 + R_L)}} \tag{3.39}$$

Equation (3.39) is an optimized current for the load. The optimization condition is $k_{12}^2Q_1 = k_{23}^2Q_3$. And the Eq. (3.39) can be deformed to:

$$I_{Optimal} = \frac{X}{1 + 2X} \cdot \frac{U_s}{\sqrt{(R_S + R_1)(R_3 + R_L)}}, X = k_{12}k_{23} \cdot Q_2\sqrt{Q_1Q_3} \tag{3.40}$$

where X is a factor defined by us. It reflects the energy coupling strength between the primary and secondary sides. By observing the equation, it is not to find out the lower the resistances are, the higher the current on load is. And the higher the X factor is, the higher the current is. For ease of comprehension, we define two equivalent Q factors.

$$Q_{Equivalent-1} = \sqrt{Q_1 \cdot Q_2}, \; Q_{Equivalent-2} = \sqrt{Q_2 \cdot Q_3} \tag{3.41}$$

where $Q_{Equivalent\text{-}1}$ and $Q_{Equivalent\text{-}2}$ are the equivalent Q factors for stand-alone primary and secondary sides. Meanwhile, we define an equivalent coupling factor.

$$k_{Equivalent} = \sqrt{k_{21} \cdot k_{23}} \tag{3.42}$$

Using the two equivalent Q factors and the equivalent coupling factor, designers could assume there is no middle resonator. Therefore, all equations and optimization methods for the LC-pair can be applied to structure of the Multiple-resonators.

The Multiple-resonators can be adopted in many biomedical applications. The purposes may include higher power transfer efficiency over a certain distance or longer transfer distance with certain transfer efficiency.

Here, we introduce a design for disposable biomedical microsystems. Actually, many biomedical microsystems suffer from the risk of cross infection. As a result, most biomedical applications use disposable devices for patients. However, although each patient touches only single device, the doctors and especially the nurses have to contact with many used devices. This phenomenon causes the increased risk of the cross infection of nurses, doctors, and even the patients. Suppose a used biomedical device is required to be processed. Maybe data is needed to read out. The best way to avoid cross infection is a non-contact solution, which may the wireless power transfer and the wireless data connectivity. Therefore, we proposed an experimented prototyping system as shown in Fig. 3.16.

In the above figure, two clips connect to an AC power source. The loaded inductor in the primary circuit is on the left side, and a ferrite core is used to adjust the inductance of the loaded inductor. A tuning capacitor is in the middle.

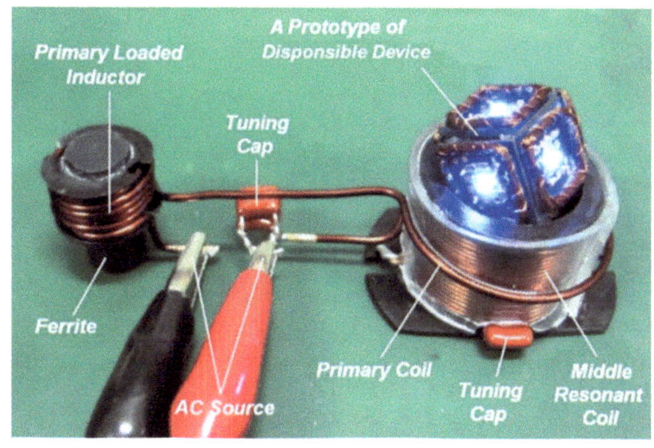

Fig. 3.16 A prototype of the Multiple-resonators for disposable biomedical microsystems

The primary coil on the right side is actually a single turn coil and is 33 mm in diameter. The single-layer multiple-turn middle resonant coil on the right side is 30 mm in diameter. Tuning capacitor is adopted to keep it resonant. The disposable secondary side is fabricated as a cube-like device. The dimension of one side of the cube is 13 mm. Six orthogonally coils are employed on the cube to ensure the energy can be successfully transferred from any direction. In contrast with those applications having fixed position and orientation between the primary and the secondary coil, our design for the secondary side emphasizes it is an omni-directional wireless power receiver, so the nurse could just place the disposable device on the top of the middle resonator without any attention to its posture. This feature is very useful for the biomedical devices since they are usually very small and in irregular shapes.

Figure 3.17 shows a close photo of the secondary side. There are tuning capacitors and LEDs on the cube. The power recovered from the six secondary coils is combined together to power the LEDs. The LEDs are the load of the transfer system. To verify the advantage of the transfer structure of the Multiple-resonator, two experiments were conducted to compare the transfer efficiencies of the regular LC-pair and the Multiple-resonator. As shown in the Fig. 3.17, the first experiment removes the original primary side. The middle resonator is connected to the AC source and acts as the primary side. The second experiment tests the complete structure of the Multiple-resonator. According to measured results, the power transfer efficiency in the second experiment is 53 % and it is 13 % higher than the efficiency in the first experiment. In the experiment, we successfully transferred power of 200 mW to the secondary side, which could cover the power demands of most biomedical microsystems.

To conclude this section, we have introduced the transfer structure of the Multiple-resonators. By inserting the middle resonator between the primary and secondary sides, it may achieve higher transfer efficiency, especially when the transfer distance is relatively long. In this section, we have presented the theoretical analysis for the Multiple-resonators. By using the defined equivalent Q factors and the equivalent coupling factor, the Multiple-resonators can be designed

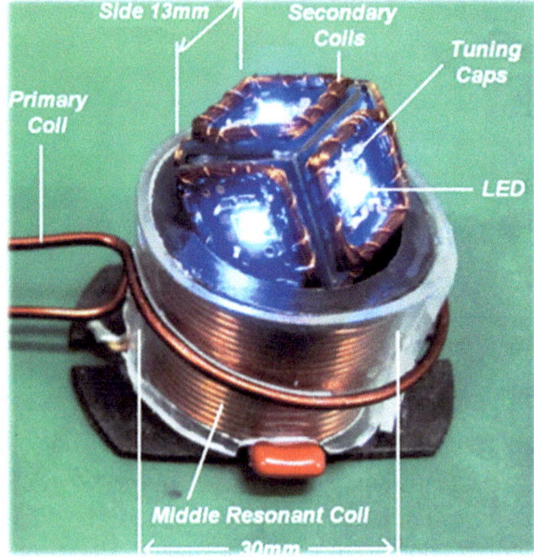

Coupling experiment from the middle resonator to a test coil

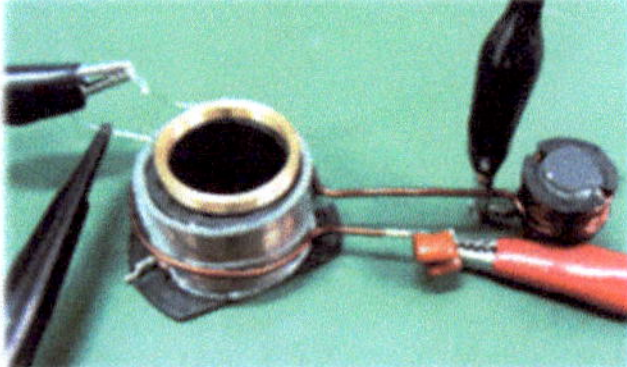

Coupling experiment from the primary coil to a test coil

Fig. 3.17 Experiments on the LC-pair and the multiple-resonators

and optimized by using methods introduced in the section of the LC-pair. Moreover, the prototype of the Multiple-resonators has been shown. Corresponding experiments have been conducted to verify the performance. The power of 200 mW has been successfully transferred to the secondary with the power efficiency improvement by 13 %.

3.4 Quad-Loops

3.4.1 Introduction

We have presented two transfer structures. In this section, we present the third structure, the Quad-loops. The Quad-loops adopts two resonator antennas and two non-resonant antennas to transfer power as Fig. 3.18 shows. The Quad-loops is original proposed in MIT's research [8].

In Fig. 3.18, there are total four power antennas. The first power antenna (Coil A) and the forth power antenna (Coil B) are non-resonators. Typically they are only single turn loops. The second power antenna is named the Coil S, which is short for source coil. The third power antenna is named the Coil D, which is short for device coil. Both of the second and the third antennas are tuned at a designed operating frequency. It's noted that the biggest difference between the previous Multiple-resonators and the Quad-loops is that there are two non-resonant

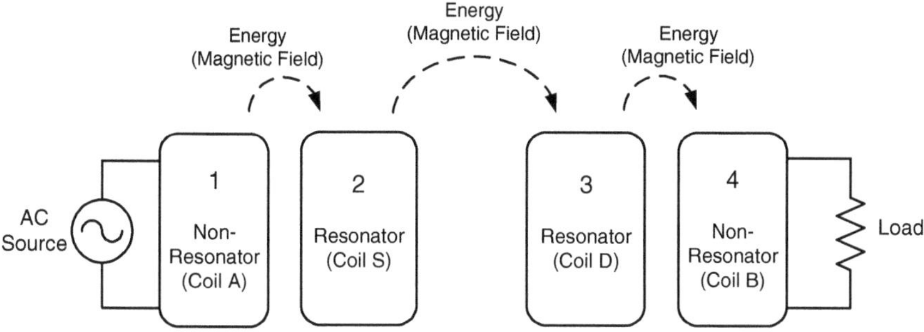

Fig. 3.18 The Quad-loops

antennas in the Quad-loops. The main function of the two non-resonant antennas is to exchange energy between the circuits and the two resonators in the middle.

The Quad-loops has two advantages and one disadvantage. The first advantage is that it's very convenient for designers to match impedance. By adjusting the separation distance or the relative angle between the Coil A and the Coil S, designers could easily change the equivalent impedance of the primary sides. By adjusting the separation distance or the relative angle between the Coil D and the Coil B, designers could easily change the equivalent impedance of the secondary sides. The second advantage is that the second and the third coils are physically set free from the circuits. Accordingly, more types of the resonators can be applied to the transfer structure. For example, we firstly introduce a design using open-ends helixes [8], and secondly present a design based on close-loop LC resonant circuits [31]. The main disadvantage of the Quad-loops is it occupies much more space than any other transfer structures, especially when the second and the third coils are the large-size open-ends helixes.

3.4.2 Detailed Designs

(a) Design based on large-size high-Q helical antennas

In 2007, a research of the wireless power transfer via strongly coupled magnetic resonances was published [8]. In their research, a Quad-loops transfer structure was designed based on 4 power antennas, in which there are two single turn loop and two large-size high-Q helical antennas. The transfer structure is shown in Fig. 3.19.

In this transfer structure, Coil A and B is a single turn loop of radius 25 cm [8]. The AC source directly connects to the Coil A. And the Load directly connects to the Coil B. The second and the third power antennas are open-end helixes of radius 30 cm and turns 5.25 [8]. The function of the Coil A and B is to make connections between the helix and the AC source or the load. The real components that transfer energy are the helixes. Using this design, it was able to transfer 60 W with 40 %

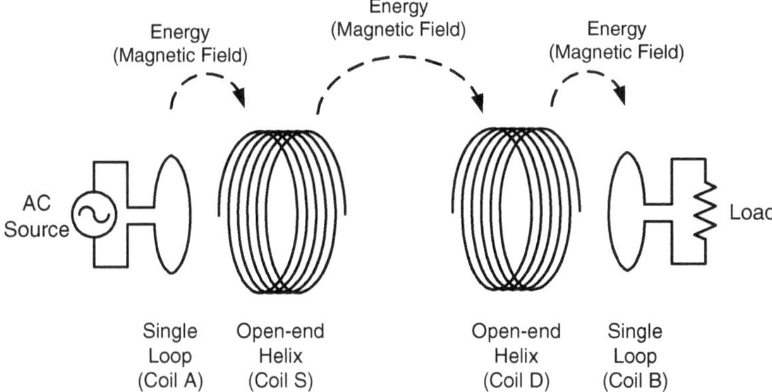

Fig. 3.19 The Quad-loops based on open-end helixes

efficiency over distance over 2 m [8]. This result is much higher than previous researches. Accordingly, it inspires many researches [19–23] to pay attention to this area. Let's firstly present the working principle of the helix, and secondly explain why the transfer structure achieves such high power transfer efficiency.

The working principles of the helix or the helical antenna can be explained by the standing wave formed by the superposition of the traveling and reflecting currents in the open-ends helix wire. Suppose a current is injected from one end of a helix wire. It travels along the wire until arrives at the other end. Because the end of the wire is an open end, the current has to reflect and travel back. As a result, there are two currents in the helix wire. They travel in opposite directions, but they have the equal frequency, the traveling speed (speed of EM wave in metal), and the current magnitude. Figures 3.20 and 3.21 shows the traveling current, the reflecting current, and the formed standing current waves in the helix.

If the raveling and the reflecting currents in the above figures are donated by the following equations:

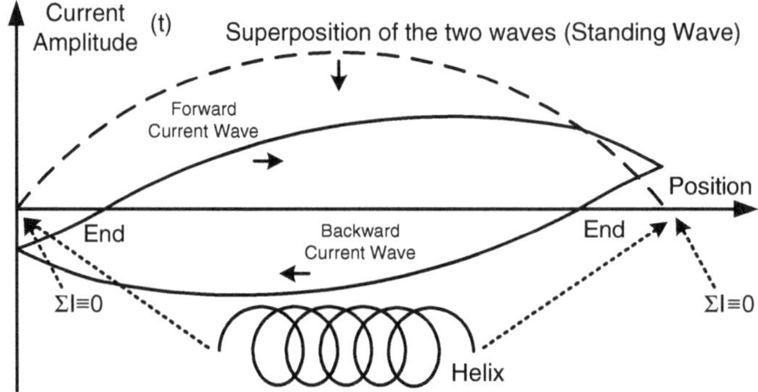

Fig. 3.20 The currents in the open-end helix at time (t)

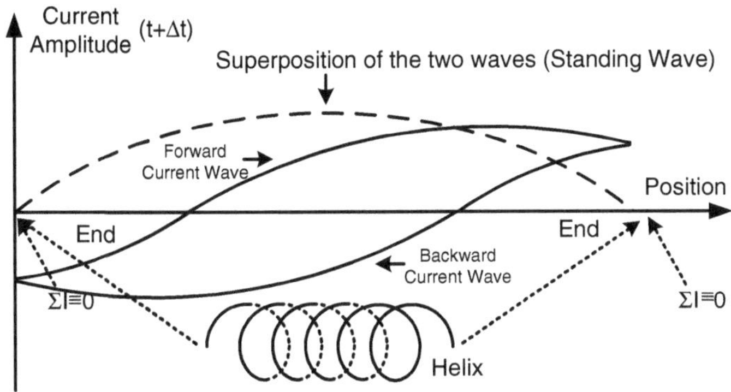

Fig. 3.21 The currents in the open-end helix at time (t + Δt)

$$\begin{cases} Y_{Forward} = A \cdot sin(\frac{2\pi}{\lambda}x - \omega t) & 0<x<l \\ Y_{Backward} = A \cdot sin(\frac{2\pi}{\lambda}x + \omega t) & 0<x<l \end{cases} \tag{3.43}$$

where, A is the amplitude of the wave, ω is the angular frequency, measured in radians per second, is the wavelength in meters, and x and t are variables for longitudinal position and time, respectively. The superposition of the two current can be given by:

$$Y_{Superpositon} = 2A \cdot cos(\omega t) \cdot sin(\frac{2\pi}{\lambda}x) \quad 0<x<l \tag{3.44}$$

The segmented line in two above figures shows the superposition of the two waves. Obviously, if the length of the helix wire equals $\lambda/2, \lambda, 3\lambda/2$, and so on, the standing wave could appear in the helix. In other worlds, at some specific frequencies associated with the total length of the wire, the two currents traveling in opposite directions would produce a standing wave. The current standing wave would build up an oscillating electromagnetic field and emit nonradiative power out of the helical antenna. The helix wire becomes a nonradiative electrometric resonator. Because there is ohmic loss in the metal wire and radiation loss in space, power is needed to be supplemented continuously.

The greatest advantage of the design is that it is capable of transferring wireless power over a distance of several meters. The transfer efficiency of the design is achieved by the high Q factor of the helixes. According to the previous research [8], the theoretical Q factor for the helix is estimated to be around 2,500 and the measured value is 950. The result is much higher than the regular LC resonators. The regular LC resonators typical have the Q factors of around 100. According to previous Eq. (3.19), Fig. 3.22 presents the difference of the transfer efficiency caused by the Q factors.

According to the above figure, the high Q factor has significant contribution to the transfer efficiency. When the coupling factor k is 0.01, the transfer efficiency of the helix resonator with the Q factor of 950 is almost 4 times higher than the

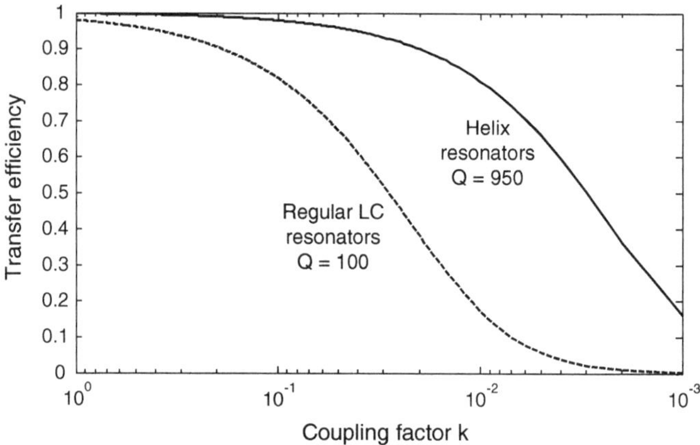

Fig. 3.22 The transfer efficiency with regards to the Q factor

efficiency of the LC resonator with the Q factor of 100. This is the essential reason that the transfer structure of the Quad-loops using the open-ends helixes has a very higher efficiency.

The main weakness of the design is that it requires large power antenna. The diameter in the above research is 30 cm [8]. It's too large for most biomedical microsystems. As a result, most biomedical applications won't adopt this transfer structure. However, we will found this transfer structure is very useful in some special applications introduced in later chapters. It can be used to efficiently transfer energy from position-fixed equipments to a jacket on a patient. The jacket would repeat the energy and deliver it to the implants in human body.

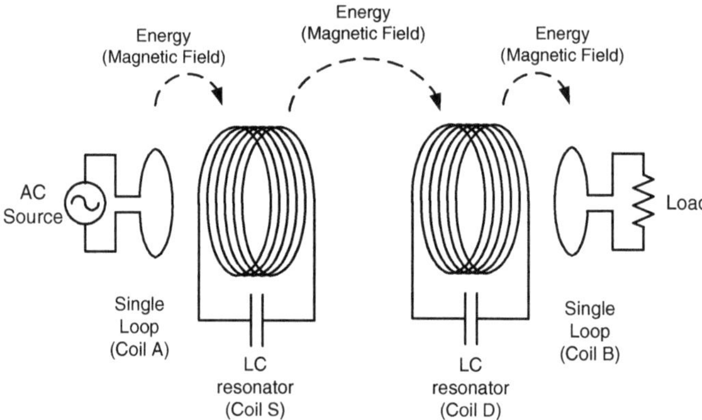

Fig. 3.23 The Quad-loops based on open-end helixes

(b) Design based on small-size low-Q LC resonators

The resonators in the Quad-loops can LC resonators instead of the helixes. As Fig. 3.23 shows, the Coil S and D are fulfilled by LC resonant circuits. They are tuned at the same operating frequency. The Coil A and B are still the same non-resonant single turn loops.

By changing the helixes to the LC circuits, the size of the antennas can be significantly reduced. For example, in previous research [32], the diameter of the secondary side is merely 1.28 cm. Therefore, this transfer structure can be adopted in the biomedical microsystems. However, since the helixes have much higher Q factors than the LC circuits, the transfer efficiency of the LC circuits significantly degrades. As a result, the advantage of the transfer structure is weakened. The only advantage left is the convenience of the impedance matching.

3.5 Helix-Derivatives

3.5.1 Introduction

According to previous introduction, the helix wire can be adopted as electromagnetic resonator and it promotes the power transfer efficiency by having much higher Q factor than the regular LC resonators. As a result, many works [6, 22, 23] devoted efforts to related researches. The corresponding proposed transfer structures are all named as the Helix-derivatives in the book. To improve the transfer performances, our research team also proposed several transfer structures [33] in the type of the Helix-derivatives.

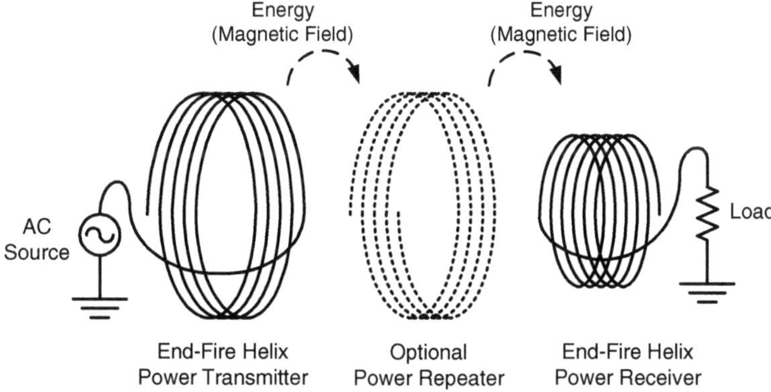

Fig. 3.24 Helix-derivatives subtype 1

3.5.2 Detailed Designs

Since the high Q factor helixes are the real components that promote the transfer efficiency, a transfer structure is proposed in Fig. 3.24, in which all resonators are the high Q factor helixes.

As the figure shows, it consists of three parts. The transmitting part (on the left) adopts an AC source and an end-fire helical antenna resonating at a specified frequency. The second part is the receiving part (on the right), which also deploys an end-fire helical antenna resonating at the same operating frequency. Essentially speaking, it is proposed to electrically excite the helix loops instead of magnetically excite them like conventional structures [8]. There is another optional part (in the middle). It's made of one or multiple optional power repeaters. They help to deliver the energy to farer space. It is noted that the structure can be asymmetrical. The receiving helix may have smaller diameter than the transmitter, so it can be adopted in power receivers with limitations on size. Although the receiving helix may have smaller diameter, the diameter is still as large as dozens of centimeters. Considering that most biomedical applications, like endoscopic capsule and simulators, require a size as smaller as several centimeters, we propose another type of the Helix-derivatives as Fig. 3.25 shows.

In Fig. 3.25, the helical antenna in the receiving part is replaced by a LC resonator that could have a much smaller size under the same operating frequency. Due to the LC resonator's smaller size and the relative lower quality factor, both the coupling factor and the transfer efficiency would degrade. It can be viewed as a tradeoff between the receiver's size and the power efficiency for ultra small size applications. It can be also viewed as a mixed transfer structure of the strong-couplings and the loose-couplings.

According to the essential work principles of the helix, there could be more derived structures based on helical dipoles as shown in Figs. 3.26 and 3.27.

Figure 3.26 shows two power transmitters. The first transmitter uses an AC source. And the second transmitter uses a DC source and an H-bridge to convert

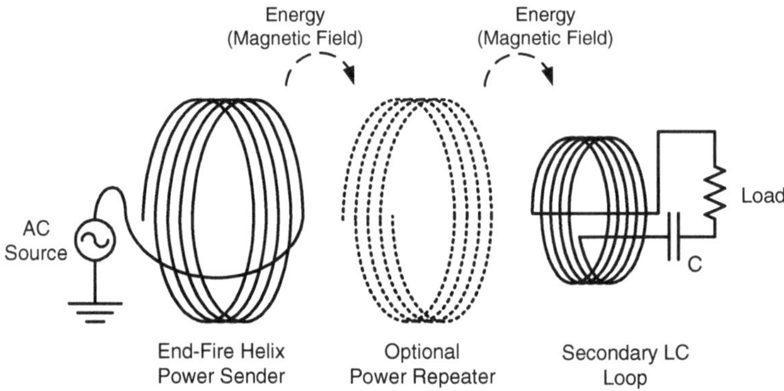

Fig. 3.25 Helix-derivatives subtype 2

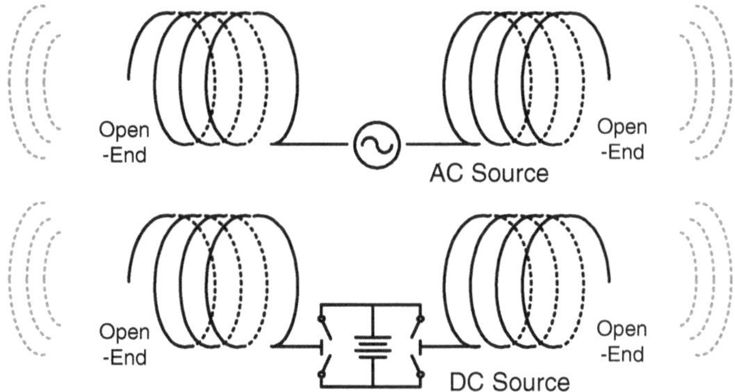

Fig. 3.26 Helix-derivatives subtype 3

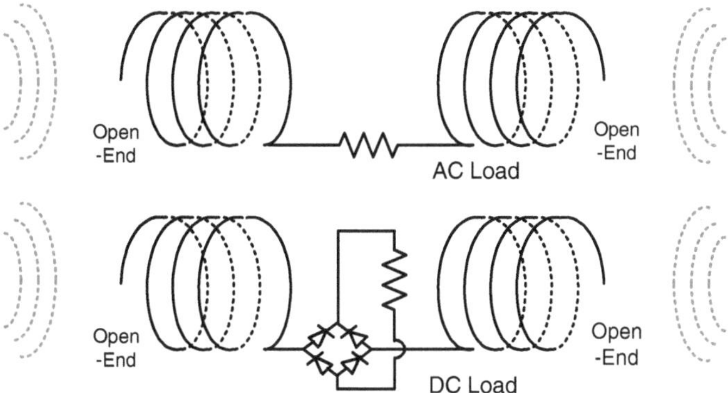

Fig. 3.27 Helix-derivatives subtype 4

the DC source to an AC source. Figure 3.27 shows to power receivers. The first receiver provides AC power to a load. The second receiver provides DC power to a load by using a full-bridge rectifier.

The four transfer structures in above figures transmit or receive energy in two adverse directions. It means they are not suitable for the point-to-point power transfer. However, they may help the applications requiring omnidirectional transfer. For example a wireless power transmitting station for multiple receivers.

In the previous section, we have explained the working principle of the helix by using the theory of the standing wave. The working principle of the helix can be also explained by using microscopic physics. Suppose positive electrons travel along a helix wire. When they travel near to one end, the voltage of the end rises because of the accumulation of the electrons. Consequently, positive electrons are repelled and reflected back to the other end of the helix. Once the traveling cycle of electrons matches the external stimulation frequency, electrons resonantly oscillate between the two ends. The macroscopic behavior of electrons forms oscillating current and generates oscillating magnetic field. The real working

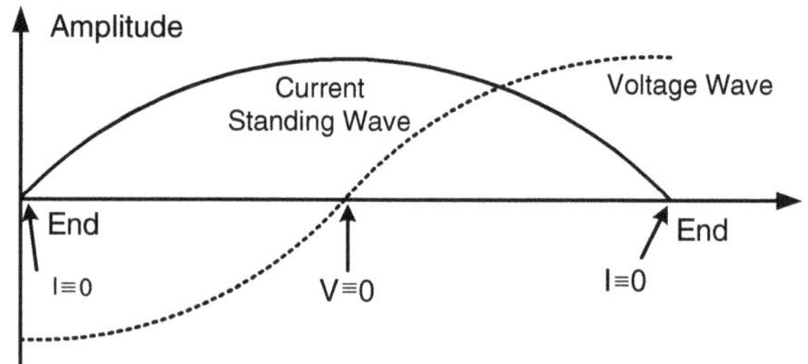

Fig. 3.28 The current and voltage distribution patterns at the first resonating frequency

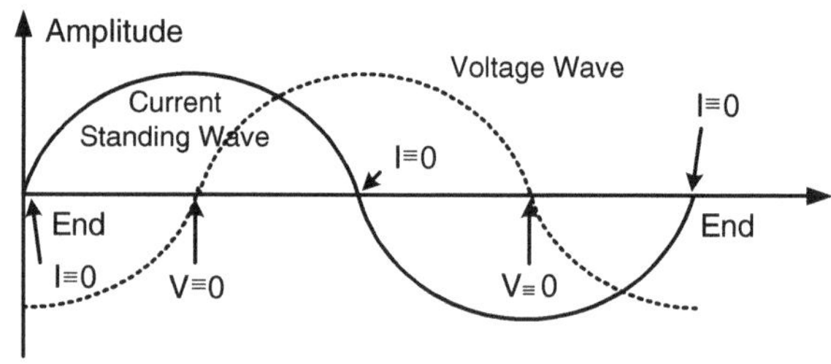

Fig. 3.29 The current and voltage distribution patterns at the second resonating frequency

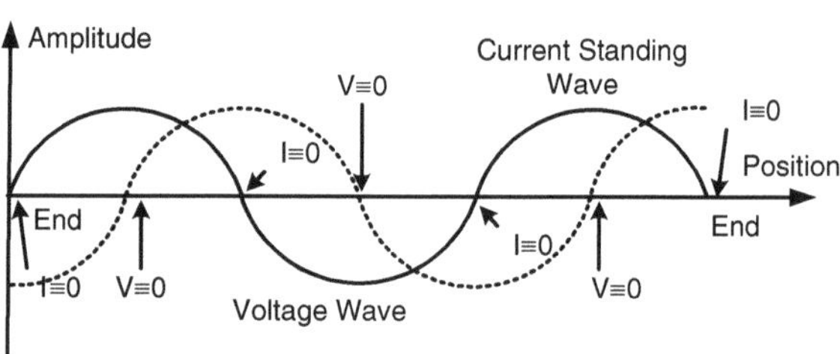

Fig. 3.30 The current and voltage distribution patterns at the third resonating frequency

principle of the helix wire is much more complex than the explanation because the inductive coupling between the turns of the helix wire is neglected.

According to the former analysis, the distribution pattern of the current and voltage in the helical antenna can be simplified and represented by sine waves. Figures 3.28, 3.29, and 3.30 shows threes distribution patterns.

As Figs. 3.28–3.30 shows, since the helical antenna has a serial of natural resonating frequencies, it can be used to transfer energy at a group of resonant frequencies. The distribution pattern in Fig. 3.28 corresponds to the first resonant frequency, which is also called the basic resonant frequency. The patterns in Figs. 3.28 and 3.29 correspond to the second and the third resonant frequencies respectively.

In order to verify the transfer structure, experiments are conducted. As Fig. 3.31 shows, two helical antennas were designed. In order to keep certain distance with other objects, two tripods were used to hang the helixes in the air in our

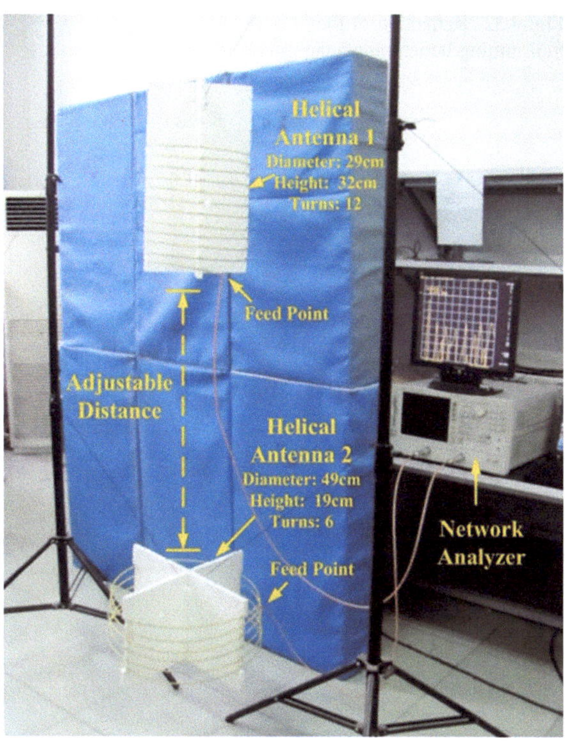

Fig. 3.31 Experiments on Helix-derivatives subtype 1

experiment. The separate distance was adjustable. An Agilent 8753ES Network Analyzer was used to measure the s-parameters in the frequency range from 1 to 100 MHz.

In the experiment, we adopt two helixes with different diameters. The diameter of the transmitter at the button side is 49 cm. The diameter of the receiving antenna of the receiver at the top side is 29 cm. Their heights are 19 and 32 cm. Their turns are 6 and 12 respectively. Both of the helixes are made of solid copper wire with the section diameter of 1.8 mm. Figure 3.32 shows the reflection of the transmitting helix.

Figure 3.33 shows the reflection of the receiving helix.

According to above figures, both of the helical antennas have multiple natural resonant frequencies. It's noted that the two helixes have the same first resonant frequency of 8.4 MHz. Therefore, the power can be transferred at this frequency. Figure 3.34 shows the measured transmission. According to the Fig. 3.34, the wireless power can be transfer at a serial of frequencies associated with the antennas' natural resonating frequencies. At the first resonant frequency of 8.4 MHz, the measured transfer efficiency is −3.91 dB over a separate distance of 0.5 m. When the transfer distance is increased to 1 m, the measured transmission efficiency degrades to −4.48 dB.

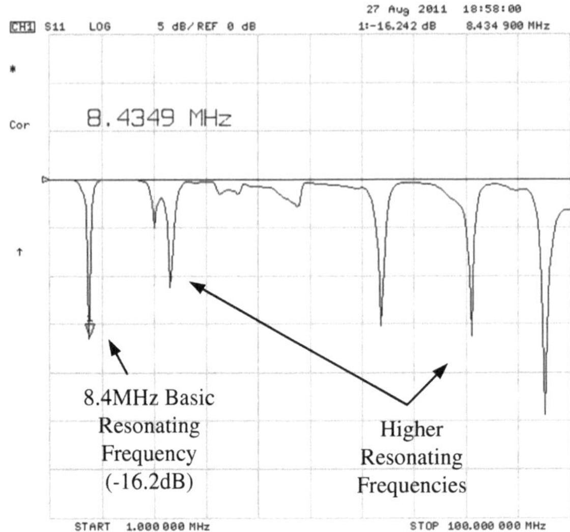

Fig. 3.32 Reflection of the transmitting helical antenna

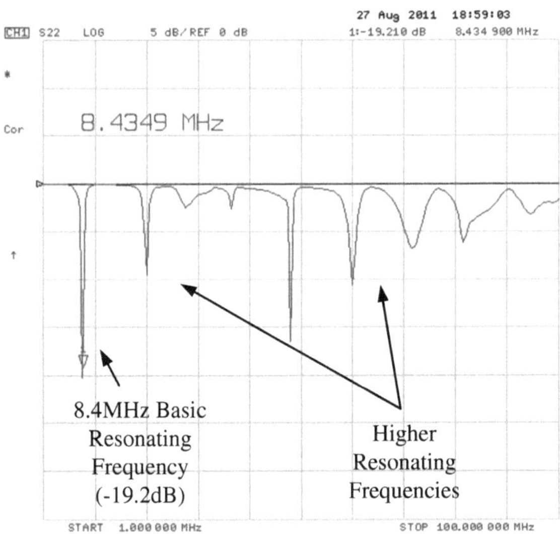

Fig. 3.33 Reflection of receiving helical antenna

To sum up, we have introduced the motivation to design the transfer structures of the Helix-derivatives. By using high Q factor helixes in the structures, the power transfer efficiency is promoted. Totally 4 subtypes of the Helix-derivatives structures are presented in this section. The transfer structure subtype 1 of the Helix-derivatives uses double helixes to transfer energy between large-size helical

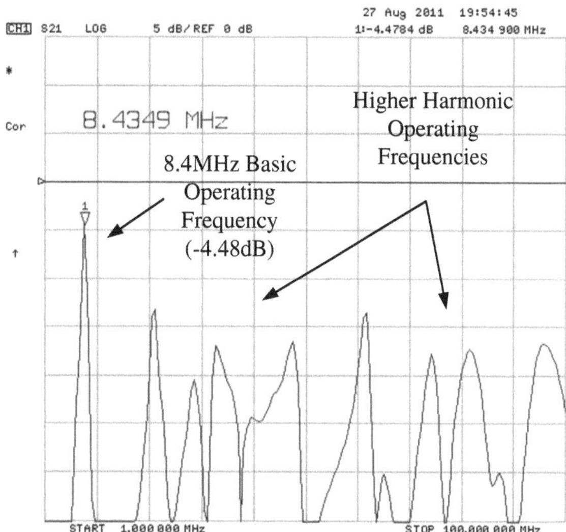

Fig. 3.34 Transmission over a distance of 1 m

antennas. To reduce the secondary side, the structure subtype 2 replaces the helical antenna at the secondary side by a LC resonant circuit. The structure subtype 3 and 4 are both proposed for omnidirectional transfers or for systems with multiple receivers. Experiments have been introduced to show the performance of the Helix-derivatives. The transfer efficiency of −4.48 dB over the separate distance of 1 m is observed in the transfer structure subtype 1.

3.6 Summing Up

To sum up this chapter, we have presented an overview of the power antennas in the wireless power transfer. Five design considerations have been illustrated for biomedical applications. Three classification methods have been discussed, and we have finally selected the classification method by making a distinction between the power transfer structures as the clue of this chapter.

According to classification method of the transfer structure, we have introduced 4 types of the structures. They are the LC-pair, the Multiple-resonators, Quad-loops, and the Helix-derivatives. Although the four structures share the same basic physical principle of magnetic coupling, they have entirely totally difficult performances and applied conditions. The four structures and their sub-type structures are summarized in Fig. 3.35.

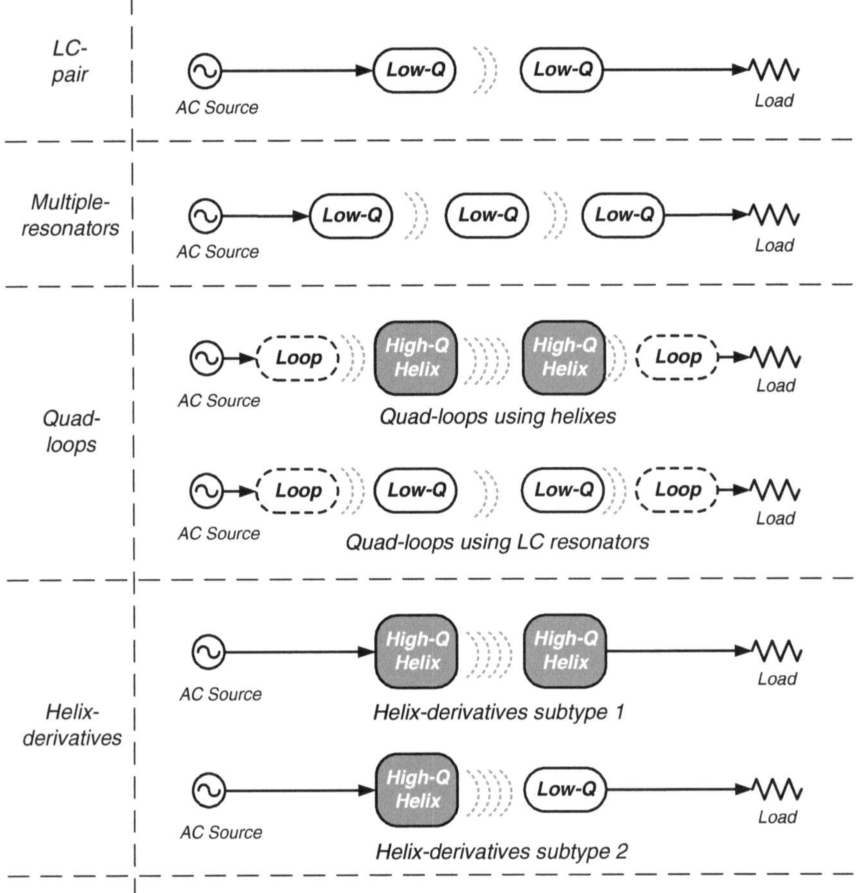

Fig. 3.35 Summary of the wireless power transfer structure in this chapter

1. LC-pair

The LC-pair is the simplest and the most frequently adopted transfer structure. Using double inductor-capacitor (LC) resonant circuit, the energy is coupled from the primary side to the secondary side. The LC-pair efficiently couples energy over relatively short distance. However, when the transfer distance increases, the efficiency significantly degrades due to the Q factor of the LC resonator is not high enough. As a result, the LC-pair can only be adopted in short range transfer. Additionally, since the LC resonator has relatively small size, it can be used in power receivers with strict size limitation.

2. Multiple-resonators

To improve the power efficiency, the transfer structure of the Multiple-resonators is proposed. By inserting additional resonators between the primary and secondary sides, the coupling strength between the source and the load is strengthened, especially when the original transfer distance is relatively long. The disadvantage of the Multiple-resonators is it occupies additional space.

3. Quad-loops

The transfer structure of the Quad-loops is very suitable for relatively long distance transfer, like decimeters or even meters. The essential advantages are the high Q factors of the helixes and the convenience of impedance matching. The problem of the Quad-loops is the helix are typical very large, like in diameter of centimeters. As a consequence, it can only be adopted to transfer energy from position-fixed equipment to jacket wear on patient, but not to implantable micro-devices.

4. Helix-derivatives

In order to reduce the antenna's size and the design complexity of the Quad-loops, the Helix-derivatives is proposed. By using double helixes, the energy is efficiently transferred from the source to the load. By replacing the second helix to a LC resonant circuit, the secondary side becomes small enough to be integrated into the implantable micro-devices.

References

1. Ghovanloo, M., & Najafi, K. (2004). Fully integrated wideband high-current rectifiers for inductively powered devices. *IEEE Journal of Solid-State Circuits, 39*(11), 1976–1984.
2. Yoo, J., Yan, L., Lee, S., et al. (2010). A 5.2 mW self-configured wearable body sensor network controller and a 12 μW 64.9 % efficiency wirelessly powered sensor for continuous health monitoring system. *IEEE Journal of Solid-State Circuits, 45*(1), 178–188.
3. Lee, S. B., Lee, H.-M., Kiani, M., et al. (2010). An inductively powered scalable 32-channel wireless neural recording system-on-a-chip for neuroscience applications. *IEEE Transactions on Biomedical Circuits and Systems, 4*(6), 360–371.
4. Nakamoto, H., Yamazaki, D., Yamamoto, T. et al. (2006). A passive UHF RFID tag LSI with 36.6 % efficiency CMOS-Only rectifier and current-mode demodulator in 0.35 μm FeRAM Technology, *ISSCC* (pp. 1201–1210).
5. O'Driscoll, S., Poon, A., Meng, T. H. (2009). A mm-sized implantable power receiver with adaptive link compensation. *ISSCC* (pp. 294–295).
6. Imura, T., Okabe, H., Uchida, T., Hori, Y. (2009). Study on open and short end helical antennas with capacitor in series of wireless power transfer using magnetic resonant couplings. *IECON* (pp. 3848–3853).
7. Zhang F., Liu, X., Hackworth, S. A. et al. (2009). In vitro and in vivo studies on wireless powering of medical sensors and implantable devices. *LiSSA* (pp. 84–87).
8. Kurs, A., Karalis, A., Moffatt, R. et al. (2007). Wireless power transfer via strongly coupled magnetic resonances. *Science 317*(5834), 83–86.

9. O'Handley, R. C., Huang, J. K., Bono, D. C., et al. (2008). Improved Wireless, Transcutaneous Power Transmission for in vivo Applications. *IEEE Sensors Journal, 8*(1), 57–62.
10. Balanis, C. A. (1982). *Antenna theory: analysis and design/Constantine A Balanis. J.*. New York: Wiley.
11. Terman, F. E. (1943). *Radio engineer's handbook.* New York: McGraw-Hill Book Company Inc.
12. Sun, T., Xie, X., Li, G. et al. (2012). A two-hop wireless power transfer system with an efficiency-enhanced power receiver for motion-free capsule endoscopy inspection. *IEEE Transactions on Bio-Medical Engineering.*
13. Lenaerts, B., & Puers, R. (2007). An inductive power link for a wireless endoscopy. *Biosensors and Bioelectronics, 22*(7), 1390–1395.
14. Si, P., Hu, A. P., Malpas, S., et al. (2008). A frequency control method for regulating wireless power to implantable devices. *IEEE Transactions on Biomedical Circuits and Systems, 2*(1), 22–29.
15. Harrison, R. R., Watkins, P. T., Kier, R. J., et al. (2007). A low-power integrated circuit for a wireless 100-electrode neural recording system. *IEEE Journal of Solid-State Circuits, 42*(1), 123–133.
16. Stielau, O. H., & Covic, G. A. (2000). Design of loosely coupled inductive power transfer systems. *International Conference on Power System Technology, 1*, 85–90.
17. Shiba, K., Morimasa, A., & Hirano, H. (2010). Design and development of low-loss transformer for powering small implantable medical devices. *IEEE Transactions on Biomedical Circuits and Systems, 4*(2), 77–85.
18. Casanova, J. J., Low, Z. N., & Lin, J. (2009). A loosely coupled planar wireless power system for multiple receivers. *IEEE Transactions on Industrial Electronics, 56*(8), 3060–3068.
19. Mur-Miranda, J. O., Fanti, G., Yifei F. et al. (2010). Wireless power transfer using weakly coupled magnetostatic resonators. *ECCE* (pp. 4179–4186).
20. Fotopoulou, K., & Flynn, B. W. (2011). Wireless power transfer in loosely coupled links: Coil misalignment model. *IEEE Transactions on Magnetics, 47*(2), 416–430.
21. Chen, C.-J., Chu, T.-H., Lin, C.-L., et al. (2010). A study of loosely coupled coils for wireless power transfer. *IEEE Transactions on Circuits and Systems II: Express Briefs, 57*(7), 536–540.
22. Zhang, F., Hackworth, S. A., Fu, W., et al. (2011). Relay effect of wireless power transfer using strongly coupled magnetic resonances. *IEEE Transactions on Magnetics, 47*(5), 1478–1481.
23. Imura, T., Okabe, H., Uchida, T. et al. (2009). Study on open and short end helical antennas with capacitor in series of wireless power transfer using magnetic resonant couplings. *IECON* (pp. 3848–3853).
24. Ooi, B.-L., Xu, D.-X., Kooi, P.-S., et al. (2002). An improved prediction of series resistance in spiral inductor modeling with eddy-current effect. *IEEE Transactions on Microwave Theory and Techniques, 50*(9), 2202–2206.
25. Kundert, K. (2006). *Modeling Skin Effect in Inductors.* CA: Designer's Guide Consulting Inc.
26. Bartoli, M., Reatti, A., & Kazimierczuk, M. K. (1994). High-frequency models of ferrite core inductors. *IECON, 3*, 1670–1675.
27. London Metal Exchange. http://www.lme.com/.
28. Yates, D. C., Holmes, A. S., & Burdett, A. J. (2004). Optimal transmission frequency for ultralow-power short-range radio links. *IEEE Transactions on Circuits and Systems I: Regular Papers, 51*(7), 1405–1413.
29. Shiba, K., Nagato, T., Tsuji, T., et al. (2008). Energy transmission transformer for a wireless capsule endoscope: Analysis of specific absorption rate and current density in biological tissue. *IEEE Transactions on Biomedical Engineering, 55*(7), 1864–1871.
30. Sun, T., Xie, X., Li, G. et al. (2010). An asymmetric resonant coupling wireless power transmission link for Micro-Ball Endoscopy. *EMBC* (pp. 6531–6534).

31. Cannon, B. L., Hoburg, J. F., Stancil, D. D., et al. (2009). Magnetic resonant coupling as a potential means for wireless power transfer to multiple small receivers. *IEEE Transactions on Power Electronics, 24*(7), 1819–1825.
32. Kumar, A., Mirabbasi, S. & Chiao, M. (2009). Resonance-based wireless power delivery for implantable devices. *BioCAS* (pp. 25–28).
33. Sun, T., Xie, X., Li, G., Gu, Y., & Wang, Z. (2013). Indoor wireless power transfer using asymmetric directly-strong-coupling mechanism. *Microwave and Optical Technology Letters, 55*(2), 250–253.

Chapter 4
Wireless Power Converters

Abstract A wireless power transfer is composed of power antennas and power circuits. Besides the antennas introduced in the Chap. 3, many circuit techniques have already been developed in past years. They can be classified into the power converters and the power management techniques. We introduce the power converters in this chapter. First, we build a circuit model for the power converters. Second, their design considerations are discussed for biomedical applications. These considerations include higher system efficiency, better system adaptability, smaller circuit size, stronger system reliability, and so on. Third, the techniques of the converters are elaborated one by one. They are the AC–DC converters, the DC–DC converters, and the DC–AC converters. Since the design of the AC–DC converter is so important for the wireless power receiver and the technology is developing fast, we mainly talk about the AC–DC converter in this chapter.

4.1 Introduction

A wireless power transfer system is made of power antennas, power converters, and power management circuits. In the last chapter, we have introduced the power antenna techniques for biomedical microsystems. In this chapter, we present the power converters with detail.

The power converter is a circuit that converts electric energy from one from to another. In the wireless power transfers, the power converters are widely adopted to change energy between AC and DC domains, or change the voltage, the current, or the frequency. To clearly analyze the power converts, we firstly illustrate a circuit model of the converters as Fig. 4.1 shows. In this model, the up side part is the power transmitting circuit, and the down side part is the receiving circuit. The transmission is wirelessly achieved by using couplings between the antennas. Four power converters can be found in the model.

T. Sun et al., *Wireless Power Transfer for Medical Microsystems*,
DOI: 10.1007/978-1-4614-7702-0_4,

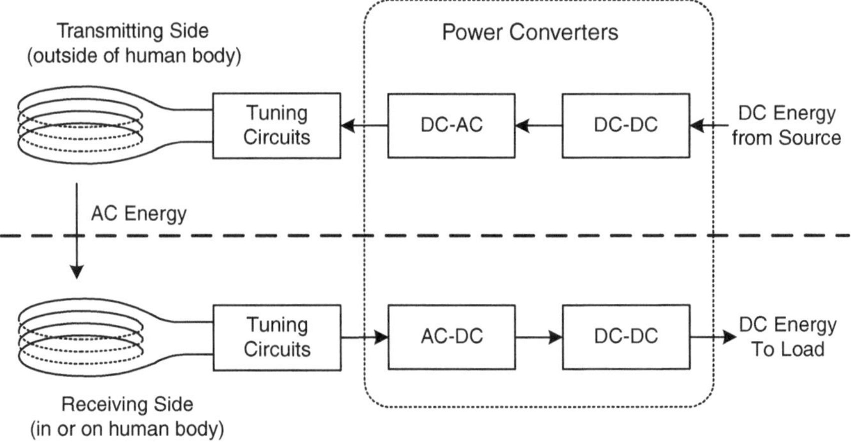

Fig. 4.1 The circuit model of power converters in the wireless power transfer

The transmitting side is commonly placed outside of human body in biomedical applications. It's composed of at least a DC energy source, a DC–DC converter, a DC–AC converter or an inverter, a tuning circuit, and a power transmitting antenna. The DC energy source can be a group of batteries or a power adapter using city power. The DC–DC converter is employed to convert the DC energy source to another DC voltage for the DC–AC converter. The DC–AC converter is the core of the transmitting circuit. It converts DC energy into AC energy at a specified operating frequency. The DC–AC converter is also called the power inverter. The power efficiency of the transmitting side is essentially decided by the efficiency of the DC–AC converter. Tuning circuit is adopted to allow the antenna to resonant at the designed operating frequency.

The receiving side is commonly placed on or in the human body. It is constitute of at least a power receiving antenna, a tuning circuit, an AC–DC converter or a rectifier, a DC–DC converter, and a load. The tuning circuit makes sure that the power receiving antenna resonates at the operating frequency. Accordingly, the antenna picks up AC magnetic energy in space. The AC–DC converter converts the AC energy into DC power. The AC–DC converter is also called the rectifier, which maybe is a full-wave rectifier, a half-wave rectifier, a switch-mode rectifier, or any other type of rectifiers. The rectifier is definitely the most important circuit in the receiving side. The power receiving efficiency is essentially determined by the rectification efficiency. Accordingly, the circuit technique for the rectifier is the key topic of this chapter. Since the output of the rectifier is unregulated DC power, DC–DC converter is adopted in the receiving side to provide regulated and stable DC power for the load.

According to the above model, it's clear that there are three types of power converters adopted in the wireless power transfer. They are the AC–DC converter or the rectifier, the DC–AC converter or the inverter, and the DC–DC converters. The DC–DC converter includes two subtypes. They are the linear regulators and the switch-mode DC–DC converters. In this chapter, we present these power converters.

4.2 AC–DC Converters

The AC–DC converter or the rectifier is the most important power converter in the wireless power transfer. It converts the AC energy received by coils to DC power for Loads. It's the key to enhance the power receiving efficiency.

The classical full-wave and half-wave rectifiers based on diodes cannot meet the requirement of the wireless power transfer for biomedical microsystems. Actually, the structure of the classical rectifiers is very simple and reliable. However, the 0.7 V dropout voltage of the diodes is too large for the wireless power transfer in low-power low-voltage biomedical applications. Even the 0.35 V dropout voltage of Schottky diodes is not small enough. The AC voltage in the receiving coil is usually in range from 1 to 3 V. As a result, the dropout voltage of the diodes considerably degrades the power efficiency. To address this problem, researchers started to design low-dropout rectifiers. Some employed the low-dropout techniques like the threshold voltage cancellation [1] and the adaptive threshold [2], while some turned to the switch-mode rectifiers [3, 4] because the on-resistance of a switch-mode transistor consumes less energy.

Figure 4.2 shows an overview of the recently proposed rectifiers [1–8]. As you can see, the rectifiers operating at relative low frequencies usually have higher power conversion efficiency. It is quite normal to design an 80 % power efficiency rectifier at 1 MHz. However, the higher the frequency is, the lower the efficiency is. Accordingly to the above figure, the best rectifying efficiency at 2.4 GHz is merely 11.3 %. It's because every circuit process has the characteristic frequency. It's notable that the rectification efficiency in the Fig. 4.2 is peak power efficiencies. The efficiency of a rectifier changes with regards to many factors, for example the load, the input voltage, the output voltage, and so on. The peak efficiency typically appears at heavy load and high voltages.

Next, we present several state-of-the-art rectifier designs. Some of these rectifiers are optimized for power conversion efficiency at very high operating frequency, while some others are optimized for ultra small size or other features.

4.2.1 Self-Synchronous Rectifier

Figure 4.3 shows a self-synchronous rectifier [6]. The rectifier is composed of double power NMOS transistors, double power PMOS transistors, and double body voltage control circuits. The four NMOS and PMOS power transistors are the main current path. In each of the half periods, one NMOS and one PMOS transistors on the diagonal are switched on. Accordingly, currents flow from Input-1 or

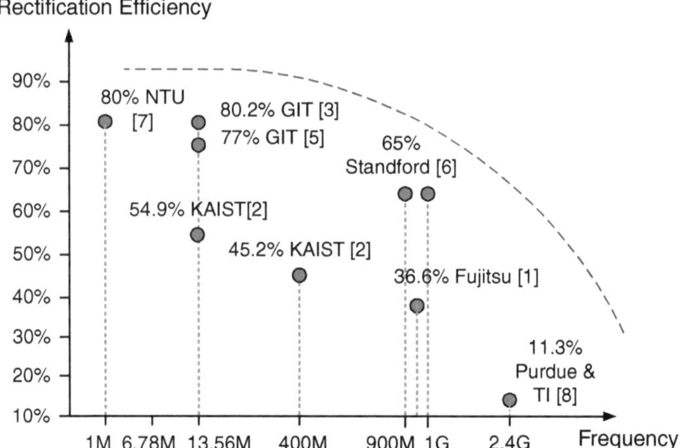

Fig. 4.2 Existing rectifiers for the wireless power transfer

Input-2 to VDD, and flow back from GND to the Input-1 or the Input-2. Multiple rectifiers of this type can be cascaded to get a higher DC voltage [6], which is called the multiple-stage rectifier. The circuit in the Fig. 4.3 can be viewed as a single-stage design.

In order to avoid latch-up, the body of PMOS transistors should be tie to the highest voltage in system, while the body of NMOS transistors should be tie to the lowest voltage. The highest voltage in this system is one of the three following options: the Input-1, the Input-2, or the VDD, while the lowest voltage in this

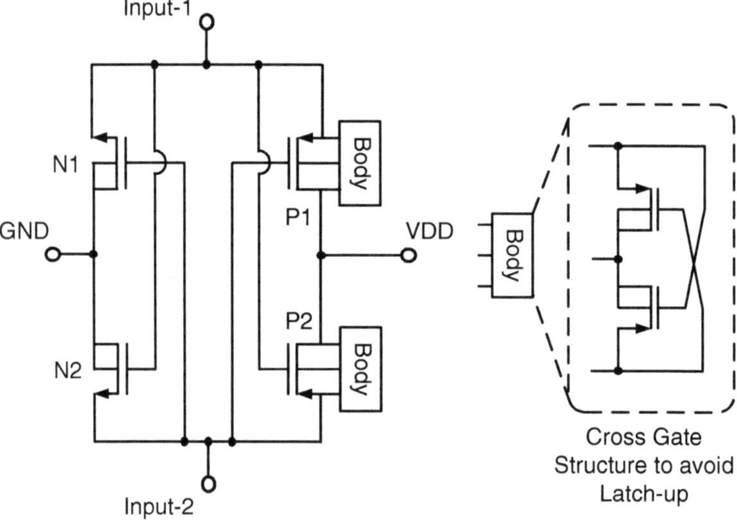

Fig. 4.3 The self-synchronous rectifier

system is one of the three options: the Input-2, the Input-2, or the GND. Therefore, a body voltage control circuit is adopted for the PMOS transistors as Fig. 4.3 shows. The body control circuit is fulfilled by two additional cross-gated PMOS transistors. It selects the highest voltage from the AC inputs and the VDD to ensure the body of PMOS transistors are tie to the highest voltage dynamically. Because the body of the NMOS transistors is actually the substrate of the whole chip, there is no way to connect the body of NMOS to the lowest voltage. Fortunately, the latch-up of the NMOS transistors won't happen.

The structure of the self-synchronous rectifier is very simple. All transistors work passively. Correspondingly, the operating frequency of this type of rectifier can be very high, like in a range from 900 MHz to 2.4 GHz. A work presented by Stanford University showed that the rectification efficiency of the rectifier reaches 65 % at the frequency of 915 MHz [6].

What is the disadvantage of the self-synchronous rectifier? The switching timing of the four power transistors is passively controlled by the input AC signals. This timing control method is quite simple but the switching time cannot be precise adjusted. The timing is actually not good enough for the improvement of the efficiency. A simulation is conducted by us and waveforms are given in Fig. 4.4 to explain the problem.

In the Fig. 4.4, there are four waveforms. They are the voltage of the VDD, the voltage of the Input-1, the current flowing from the GND to the Input-1, and the current flowing from the Input-1 to the VDD. In every period, when the voltage of the input-1 rises, the PMOS transistor is switched on and current flows from the Input-1 to the VDD. When the voltage of the input-1 falls, the PMOS transistor is switched off and the NMOS transistor is switched on. Current flows from the GND to the input-1. However, as noted in the Fig. 4.4, the rectifying currents do not flow one way but toss and turn. Because the transistors have internal resistances,

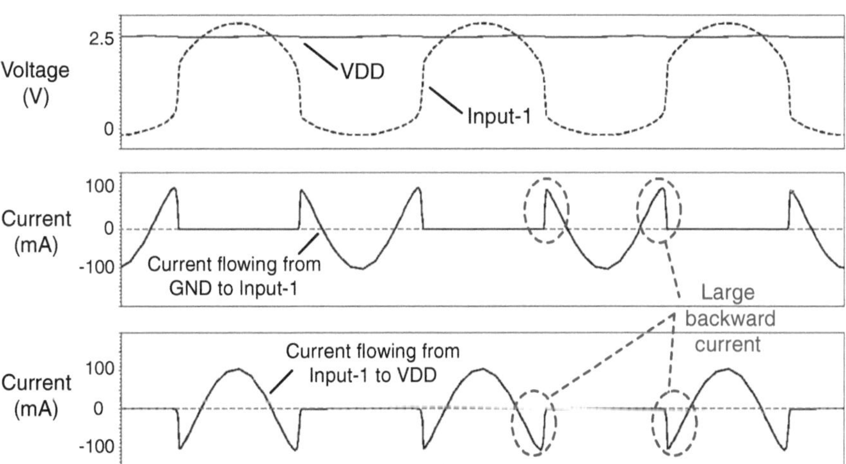

Fig. 4.4 The simulated waveforms of the self-synchronous rectifier

the existence of the backward currents decreases the power conversion efficiency. Since the circuitry has determined the duty cycle of on and off of the transistors is basically 50:50, the timing cannot be adjusted and the backward current cannot be eliminated.

As mentioned, the self-synchronous rectifier has the power efficiency of around 65 %. Consequently, the self-synchronous rectifier is not a good choice for systems working at a relative low operating frequency, like from 125 kHz to 13.56 MHz. There are some other rectifiers. They are good at working at low operating frequencies and usually have much higher power efficiencies.

4.2.2 Comparator Based Rectifier

To switch the power transistors with more precise timing and improve the power conversion efficiency, comparator based rectifiers are proposed [3, 9–11]. The idea of this type of rectifier is to use comparators to determine the switching timing for transistors. Accordingly, the problem of the backward current in the self-synchronous rectifier can be eliminated in the comparator based rectifiers. Figure 4.5 shows the circuit of the comparator based rectifier [3].

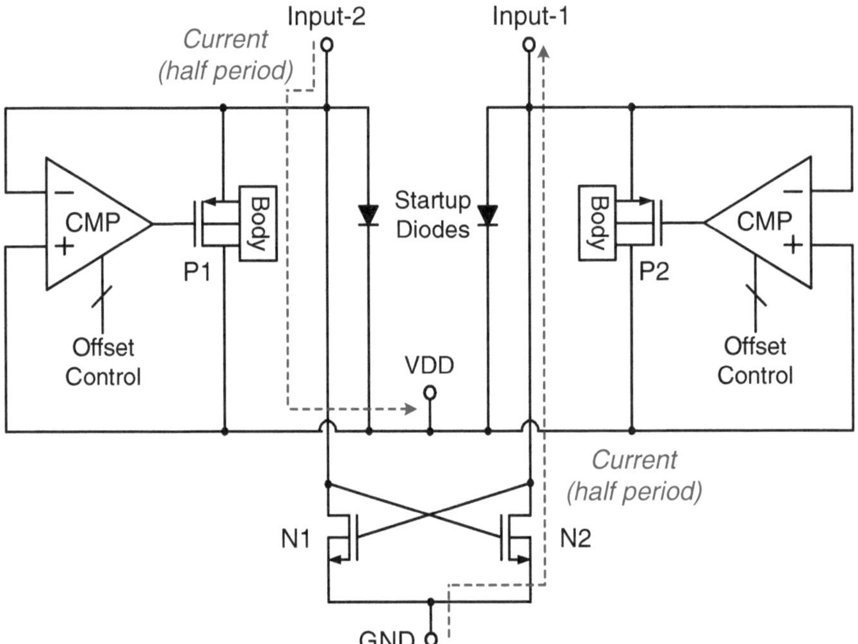

Fig. 4.5 The comparator based rectifier

The rectifier is composed of double NMOS transistors, double PMOS transistors, and double comparators. The four transistors are the main current paths. They have the large size, so they have small on resistances. The two NMOS transistors adopt cross gate structure to allow currents flowing from GND back to AC inputs. The gates of the two PMOS transistors are driven by the two comparators respectively. The negative inputs of the comparators are the AC inputs, while the positive inputs of the comparators are VDD.

Suppose the Input-1 is high voltage and the Input-2 is low voltage at this moment. If the voltage of the Input-1 is higher than the VDD, the comparator on the right site outputs low. As a result, the PMOS P2 is switched on and current flows from the Input-1 to the VDD. At the same time, due to the voltage of the Input-1 is higher than the Input-2, the NMOS N1 is switched on and current flows from the GND to the Input-2. The four transistors are switched on alternatively to constitute a full-wave rectifier. Because the voltage of the VDD is zero before the wireless power is received, two startup diodes are placed between the AC inputs and the VDD to work at the startup process. Once is the voltage of the VDD is high enough to startup the two comparators, the PMOS power transistors take over the work of the diodes. Since the two comparators have limited speed, the switching timing is actually a little bit delayed.

To explain the comparator based rectifiers more clearly, simulation is conducted by us and the waveforms are showed. In the Fig. 4.6, there are five

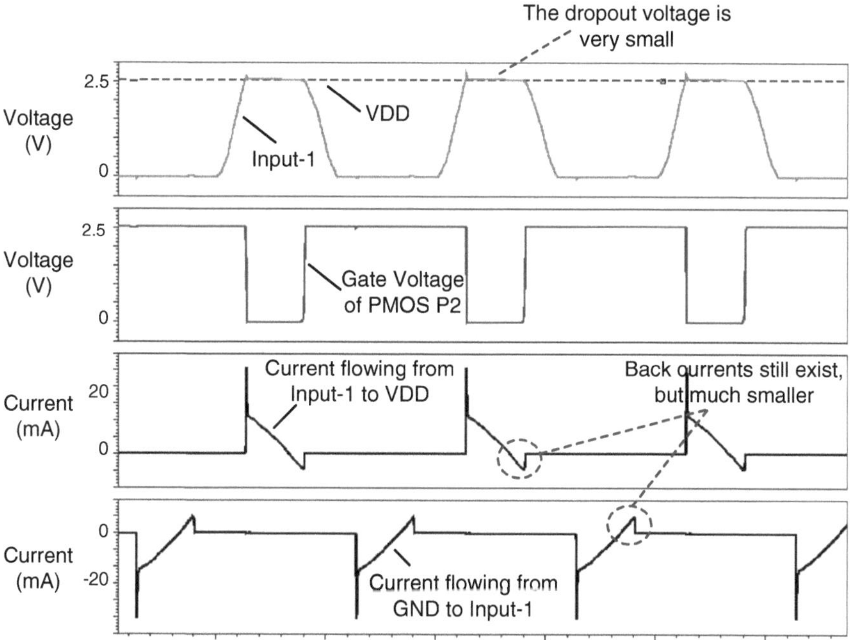

Fig. 4.6 Simulated waveforms of the comparator based rectifier

waveforms. They are the voltage of the VDD, the voltage of the Input-1, the gate voltage of the PMOS P2, the currents through the P2 and the N2 transistors. As you can see, when the Input-1 rises, the comparator at the right side outputs high and the PMOS P2 is switched on. There is only a little dropout voltage between the Input-1 and the VDD, which means the on-resistance is very low and the power efficiency is good. It's noted that the current through the PMOS P2 and the NMOS N2 transistors still have back current, but comparing to the self-synchronous rectifier introduced before, the back current is already much smaller. The back current can be alleviated by trimming the offset of the comparators as Fig. 4.5 shows.

The comparator based rectifier has good efficiency performance at low frequency band. A work by GIT reported the efficiency reaches 80.2 % at the operating frequency of 13.56 MHz [3]. The advantage of the power efficiency makes the rectifier a common choice for the wirelessly powered biomedical microsystems. The drawback of the rectifier is the speed. The delay caused by the two comparators limit the highest operating frequency of the rectifier. There are two ways to increase the speed. The first way is to increase the current of the comparators. The second way is to insert buffers or inverters between the comparators and the gates of the PMOS transistors. However, the both ways consume extra power. As a result, the comparator based rectifier is only adopted in applications that work under the frequency of 13.56 MHz.

4.2.3 Full-NMOS Full-Wave Rectifier

As introduced, the regular comparator based rectifier uses two PMOS and two NMOS transistors. Here, we propose a full-NMOS rectifier for higher power conversion efficiency since the n-channel device offers an advantage over the p-channel device. The essential reason is the n-channel device has a smaller junction area and an on-resistance. Figure 4.7 shows the proposed rectifier.

As Fig. 4.7a shows, Input-1 and Input-2 are the AC input terminal of the rectifier, and VDD and GND is the DC output terminal. There are four power NMOS transistors in main current paths. They are N1, N2, N3 and N4. The two NMOS transistors N3 and N4 on the low side use cross-gate structure as conventional circuit does, while the gates of the two NMOS transistors N1 and N2 on the high side are driven by double bootstrap circuits. Two 5 pF on-chip metal–insulator-metal capacitors are used in the bootstrap circuits. The gate of the two NMOS transistors toggles between ground and a voltage higher than the AC inputs. It ensures a low on-resistance for rectifying current. The rectifier was implemented in UMC 0.18 μm process. Accordingly, the measured PCE reaches 87 % when the operating frequency is 1 MHz and the output power is 94 mW.

The switching timing of the rectifier is controlled by two independent low-startup comparators as Fig. 4.7b shows. Since the proposed comparator needs only

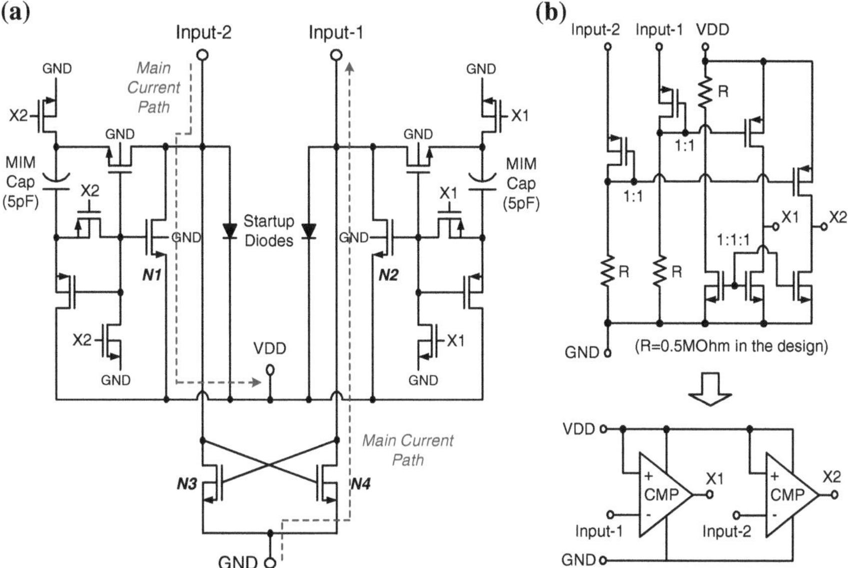

Fig. 4.7 The full-NMOS rectifier with low-startup power-lines comparators. **a** Main part of the full-NMOS full-wave rectifier. **b** Double low-startup comparator

one transistor turning on in each current branch, the startup power voltage of the comparators is only 0.7 V.

The disadvantage of the full-NMOS rectifier is it generates high voltage on the gate of the NMOS transistors. It may damage the gate oxide. Accordingly, the full-NMOS rectifier should be only used in low-voltage applications, so the gate voltage on the NMOS will not exceed the technology limitation. Fortunately, the most of biomedical microsystems are low-voltage circuits, typically in range from 3.3 to 12 V.

4.2.4 Parallel Rectifiers

The essential working principle of the wireless power transfer is the Faraday's law of induction [11]. If the transmitting coil cannot generate magnetic flux crossing the receiving coil, the transfer cannot be done. The most applications adopt a single pair of transmitting and receiving coils because their relative orientation is clear. However, in some biomedical applications, the omnidirectional wireless power transfer is needed. For example, when an endoscopic capsule is swallowed by a patient, the capsule's posture in the patient is random and hardly known. As a result, in order to ensure the wireless power can be transferred to the capsule in any capsule's posture and position, multiple coils must be used.

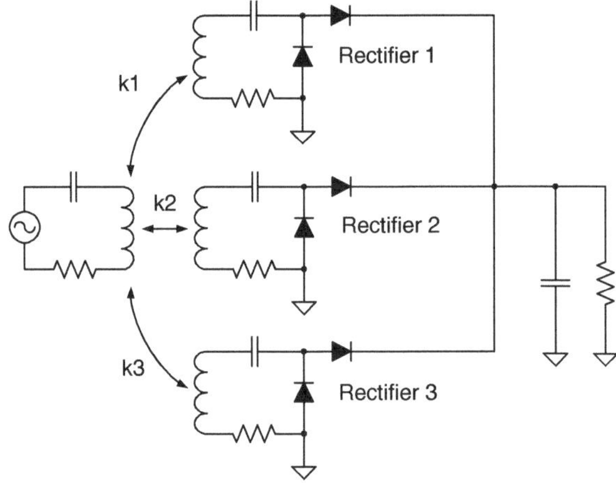

Fig. 4.8 The conventional wireless power receiving circuit with three half-wave parallel rectifiers

There are at least two ways to achieve the omnidirectional wireless power transfer. One is to use 3-dimensional transmitting coils outside the patient and use single receiving coil in the capsule. The other way is to use 3-dimensional receiving coils in the capsule and use single transmitting coil outside the patient. Because the 3-dimensional transmitting coils would occupy too much space and would be very difficult for patient to wear on, most of existing systems employ the 3-demensional receiving coils in the capsule and single transmitting coil on the patient's jacket. Accordingly, the capsule adopts a group of rectifiers to pick energy from every coil.

In existing designs [12] as the Fig. 4.8 shows, there are single power transmitting coil and three receiving coils in three orthogonal directions. Each receiving coil connects to a half-wave rectifier. The output voltages of the rectifiers are determined by the receiving coil's angle to the magnetic line. The outputs of the parallel rectifiers are all tied together to combine the received power together. Because the orientations of the three coils are orthogonal, there is at least one coil that would receive the wireless power.

There is one problem in the design. When the induced voltages in the coils have large difference, the voltage at the common output point will be determined by the rectifier with the highest DC output voltage. The other rectifiers with relative low output voltages will be back-biased. For example, if the first rectifier could independently outputs the DC voltage of 3 V and the second and the third rectifiers could output 1 V, the connected output will be determined by the first rectifier and it would be 3 V. Although energy is received by other two rectifiers, they cannot deliver energy to the load because their diodes are already back-biased. The power efficiency consequently decreases. This phenomenon is very similar to the new and old batteries are mixed together.

To explain this problem more clearly, simulations were conducted by us. The simulated circuit model is showed in the Fig. 4.9.

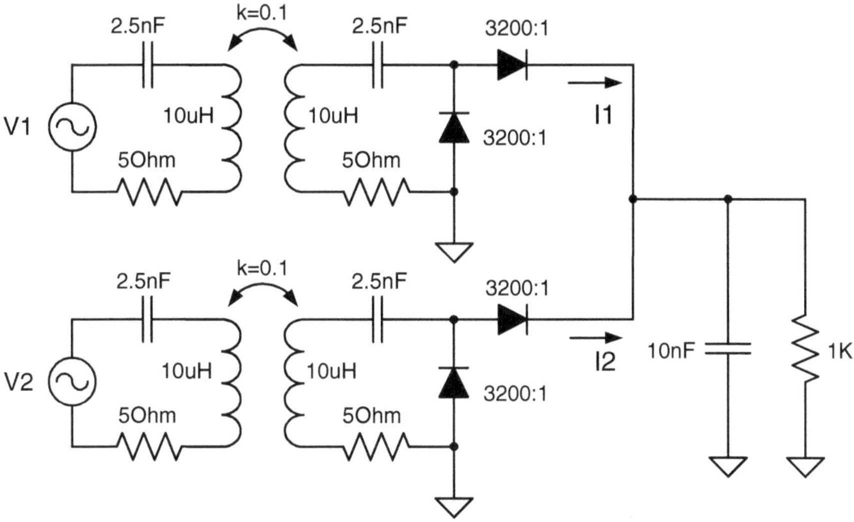

Fig. 4.9 The simulated two rectifiers and simulation conditions. **a** Simulation 1: The Simulated the currents of I1 (*up signal*) and I2 (*down signal*) when the both V1 and V2 are 1 V (V1 = V2). The result is both rectifiers provide energy to the load. **b** Simulation 2: the simulated the currents of I1 (zero like signal) and I2 (pulse like signal) when the V1 is 1 Volt but the V2 is 2 V (V2 $\gg$ V1). The result is only rectifier 2 (I2) provides energy to the load. **c** Simulation 3: the simulated the currents of I1 (*up waveform*) and I2 (*down waveform*) when both V1 and V2 are 2 V (V1 = V2). The result is both rectifiers provide energy to the load

In the Fig. 4.9, two receiving coils and two parallel rectifiers are deployed to receive and combine energy. The simulation can be also applied to triple coils and triple parallel rectifiers. In the experimented circuit, we adopted two independent AC transmitting sources, so we can transfer different power to the two receiving coils. The diodes we adopted are diodes in UMC 0.18 μm process. The outputs of the rectifiers are connected together to combine their received energy together. The received power is all delivered to a load of 1 K Ohm. In the following, we will conduct a group of simulations. The currents of the I1 and the I2 are what we are interested in, which means whether the coils can provide energy to the load or not.

As Fig. 4.10a shows, in the first simulation, the magnitudes of both the AC transmitting source V1 and V2 are 1 Volt. Accordingly, both the currents of I1 and I2 offer energy to the load. In the second simulation in the Fig. 4.10b, the magnitude of the V2 (V2 = 2 V) is set much higher than the magnitude of the V1 (V1 = 1 V). As a result, the current I1 is nearly zero, which means the coil 1 cannot provide energy into system although it has received energy. In other words, the received energy is not fully utilized. In the third simulation in the Fig. 4.10c, the magnitudes of the both V1 and V2 are 2 V. They both offer energy to the load again. According to our simulations, the resonant coil with the relatively small voltage provides little energy to the load. Actually, it is very similar to new and old

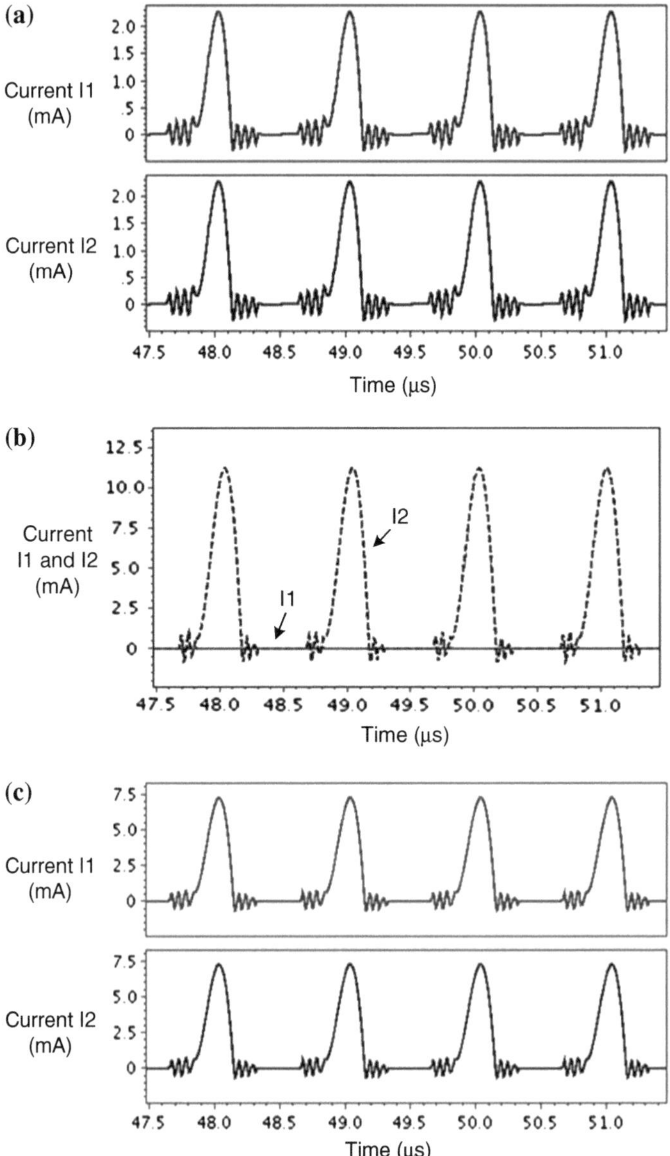

Fig. 4.10 The simulated output currents of parallel rectifiers

batteries are parallel connected. Although there is still energy in the old battery, it cannot be fully used because the output voltage is decided by the new battery.

To address the problem, a novel omnidirectional wireless power receiving circuit is proposed by our team [13]. Effectively combining energy from all directions, the proposed circuit achieves an efficiency level that is 16.2 % higher

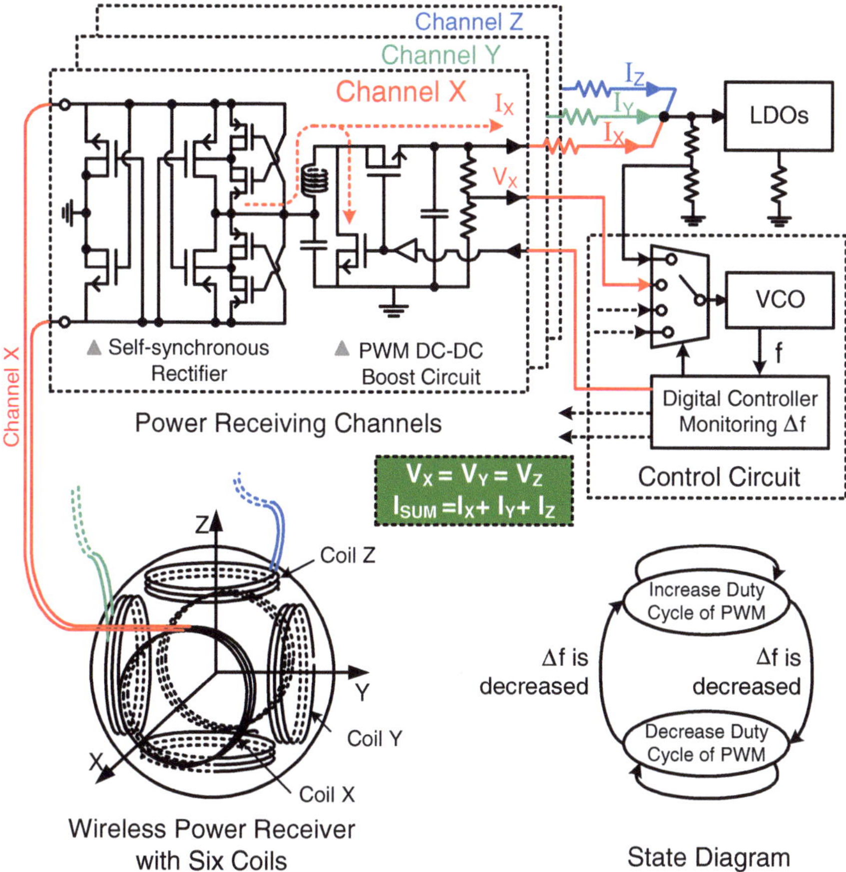

Fig. 4.11 The proposed omnidirectional wireless power receiver

than the best previous parallel rectifiers. The proposed circuit is showed in Fig. 4.11.

In the Fig. 4.11, the circuit consists of multiple receiving coils, a control unit, a group of LDO, and three receiving channels (Channel X, Y and Z), which receives wireless energy from three different directions. The circuit of each receiving channel includes a CMOS self-synchronous full-bridge rectifier and a PWM-control boost DC–DC converter. With an adjustable voltage gain control, the boost DC–DC converter converts a relatively low output voltage of the rectifier to a high level, so it's possible to guarantee for the three channels to deliver similar outputs. To keep an identical voltage output over all channels, the voltage gain of boost DC–DC converter is dynamically adjusted by the control unit. The control unit consists of three resistors, a VCO, a multiplexer, and a digital controller. The resistors are deployed at the output port of each channel. The VCO quantizes each of dropout voltages on the resistors to a value of frequency in the range from

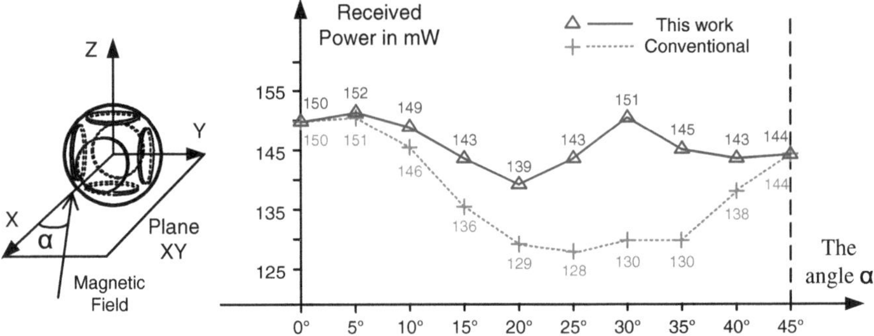

Fig. 4.12 Experiments results of the omnidirectional wireless power receiver

35.6 kHz to 1.8 MHz. The digital controller measures the frequencies, calculates the output voltage of each channel, and adjusts the gain of each boost DC–DC converter by changing the duty cycle of a corresponding PWM wave. The state diagram of the controller is illustrated in the Fig. 4.11. Consequently, the circuit enables similar output voltages over all receiving channels and effectively combines energy from all directions.

The novel receiving circuit was implemented in a 0.18 μm CMOS. In order to illustrate how the proposed circuit increases the received power, experiments are conducted. The angle between the receiver's axes X and the magnetic field is denoted as angle α. When the angle α varies, the induced voltages of channel X and Y change vary. The experiment was conducted to measure the receiving power with regards to the angle α. Only the angle α from 0° to 45° is measured because the structure of the receiving coils in our design is centrosymmetric.

The experiment results are showed in Fig. 4.12. Compared with a previous result [12], a max efficiency improvement of 16.2 % is observed at the angle α of 30°. At this angle, both channels X and Y output a reasonable large voltage but with a remarkable difference in between. In the previous work [12], the minor voltage of two channels cannot be exploited. However, our circuit effectively combines the energy from both channels. The output current of channel Y in the proposed power receiver is improved from 2.5 mA in the conventional design to 4.5 mA. The worst case of the performance improvement appears when the angle α is 0° or 45°. In fact, the received energy of the channel Y is zero when the angle α is 0°, and the output voltages of the channels X and Y equal to each other when the angle α is 45°.

4.2.5 Rectifier with Current ZCP Prediction

To further improve the power conversion efficiency, we proposed a switch-mode rectifier with ZCP prediction [14]. The rectification efficiency of this kind of rectifier can be as high as 93.6 % [14].

We firstly introduce the idea of this rectifier. For the low-power wirelessly powered implants, the dropout voltage considerably degrades the power conversion efficiency. As a result, some researches turned to the switch-mode rectifiers. Usually, the dropout voltage of the switch-mode rectifier can be as low as 0.2 V. However, there is one additional issue in the switch-mode rectifiers. That is how to determine the switching timing. A correct timing enhances the efficiency while a wrong one makes the performance worse. In former researches, there are several ways to decide the switching timing in rectifiers. The first way is the self synchronous switching [6]. In this type, the AC input signals are directly connected to the gate of transistors to drive them on or off. The second way is to adopt double comparators [3, 9, 10]. In this type, comparators are used to compare the voltages of the AC inputs and the DC output. To the best of our knowledge, the best rectifying efficiency is around 80 % [3] in present designs. To further enhance the efficiency, the switching time needs to be more precise.

Figure 4.13 shows the proposed rectifier with zero-cross-point (ZCP) prediction [14]. The essential of the switching timing of the rectification is to switch-on all forward current and switch-off all backward current. We noticed that the receiving antenna in the wireless power transfer is usually a LC resonant circuit. A fixed phase difference (pi/2) exists between the voltage and the current signals in the LC resonant circuit. Therefore, we detect the voltage ZCP ahead to predict the coming of the current ZCP behind. Accordingly, the rectifier is well prepared and switches with more precise timing.

As Fig. 4.13 shows, there are an off-chip inductor and an off-chip capacitor. When they are tuned, the voltage at Point A leads the current at Point B quarter of the cycle. The rectifier firstly delays the voltage signal at Point A and then generates digital overlapping clocks CLK1 and CLK2. The delay time ensures that the CLK1 falls slightly before the current ZCP, while the CLK2 falls slightly after the current ZCP. The two clocks are used to drive PMOS and NMOS transistors respectively. The NMOS transistor connects Point B to the ground. It keeps the backward AC current flowing in the LC circuit. The PMOS transistor connects Point B to the DC output. It transfers the forward current from the LC circuit to a load. The two transistors actually constitute a half-wave rectifier. The switching timing makes the most of the forward current flows through the PMOS and the most of the backward current flows through the NMOS. Overlapping clocks are adopted to prevent from switching on the PMOS and the NMOS transistors at the same time.

Before the startup of the rectifier, there is no DC output. So there is no power to generate the overlapping clocks and drive the PMOS and the NMOS transistors. Therefore, two bypass diodes are assisted as shown in Fig. 4.13. They are responsible to constitute a half-wave rectifier to startup the system. When the output DC Power reaches approximately 1.2 V, the overlapping clock generator and the driving stages start to work. There is little current in the diodes after the rectifier starts because the PMOS and the NMOS transistors provide lower resistance paths. To better explicit the working principles of the rectifier, waveforms are given in Fig. 4.14.

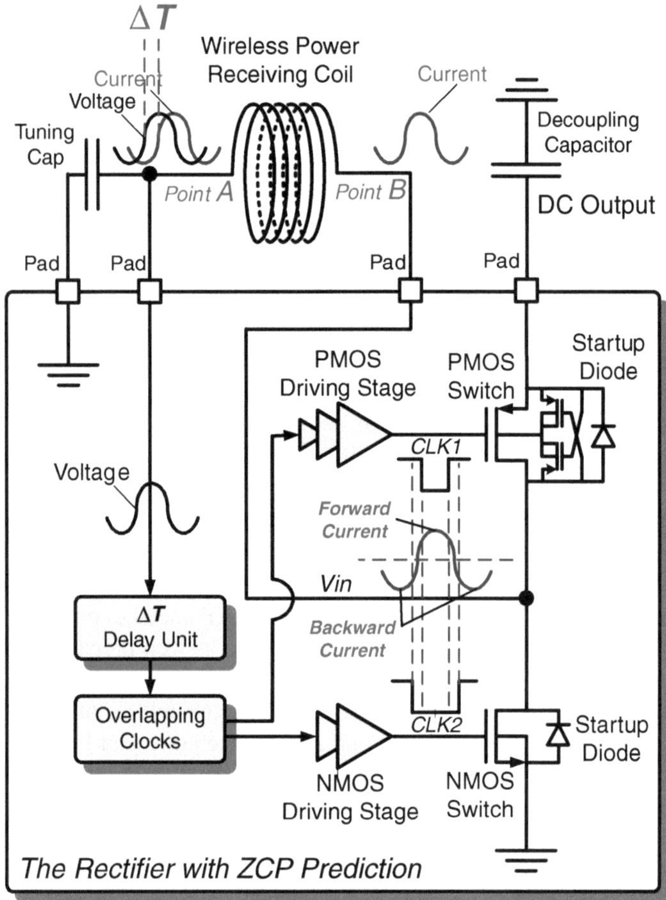

Fig. 4.13 The rectifier with zero-cross-point (ZCP) prediction

The waveforms in the above figure include the voltage of the Point A, the delayed signal of the Point A and the rectified current. The voltage of Point A is a sinusoidal wave. The rectifier delays the wave approximately 32 ns and generates two overlapping digital waves CLK1 and CLK2. They drive the PMOS and the NMOS transistors respectively. As a result, the current is precisely rectified. Figure 4.15 shows the examples of the correct current waveform and another two wrong current examples. The correct rectified current is actually the upper half of a sine wave. If the delay time is not precisely controlled, like too much or too less, negative currents would appear and the efficiency decreases. The first wrong current waveform example has negative current at the beginning of switching on, and the second wrong current waveform example has negative current before it's switched off.

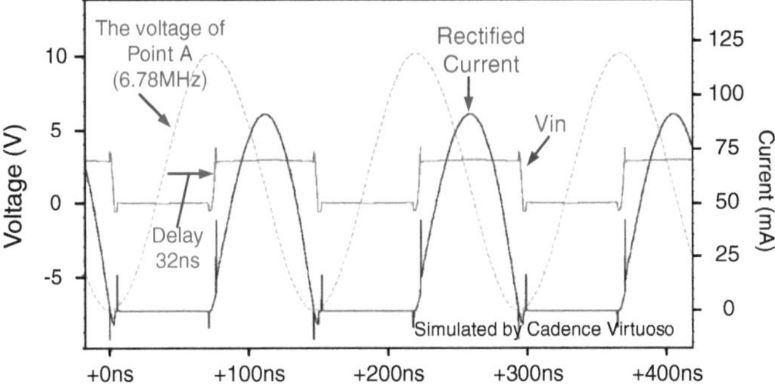

Fig. 4.14 The waveforms in the rectifier with ZCP prediction rectifier with ZCP prediction

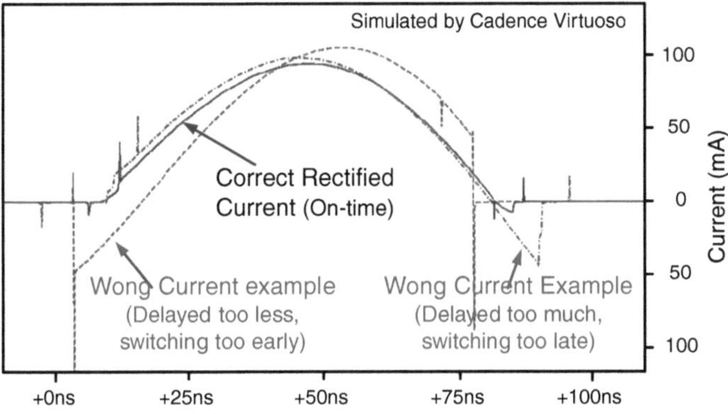

Fig. 4.15 The examples of the correct and wrong rectified currents

The proposed working principle above is simple but efficient. The key to the higher efficiency of the rectifier is it predicts but not detects the time of the current ZCP. Actually, if we just detect the current ZCP at Point B, there will be two problems. Firstly, detecting the current ZCP needs time and to drive the MOS transistors needs time. So the actual switching time will be delayed. Secondly, the transistors between the DC output and the ground are not allowed to switch simultaneously to prevent current from flowing directly from the DC to the ground. Non-overlapping is the method, but it uses delays. Consequently, these delays make the switching timing inaccurate. Unlike all previous rectifiers [1–10], the proposed rectifier makes full use of the features of the LC resonant circuit and achieves higher rectification efficiency.

As introduced, the expected delay is about a quarter of the input sine wave cycle. We adopted the classical RC delay circuit to achieve the delay unit. However, it's well known that the precision of the resistance and the capacitance values in the integrated technologies is only about 20 %, so we designed four selectable delays. We have to trim the delay by using external pads. This is very inconvenient. Therefore, we proposed another verified method to address this problem. It is the called the automatic delay control [14]. This technique does not enhance or degrade the performance. It's only designed for convenience.

The idea of the automatic delay control is to firstly check whether the rectifier is working on the perfect timing or not. If not, it adjusts the delay discreetly. Clearly, the key point is to establish a criterion that what is the perfect timing and how to check it out. The Fig. 4.16 shows our solution.

Actually as we introduced, the CLK1 should fall slightly before the current ZCP and the CLK2 should fall slightly after the current ZCP. Furthermore, the voltage at Point B has the same phase with its current. Therefore, in our solution in the Fig. 4.16, we sample the CLK1 at the rising edge of the voltage of Point B (VB for short), and sample the VB at the rising edge of the inverted voltage of the CLK2 (CLK2B for short). The sampling results reflect the timing. If it's not our desired result, the circuit is to adjust the delay unit to change the timing. Specifically, it employs two D Flip-Flop (DFF), three 2-input AND gates (AND2), and a digital state machine. The first DFF stores the value of CLK1 at the rising edge of VB and the second DFF stores the value of VB at the rising edge of CLK2B. The three following AND2 classifies the results in three types. They are 'too early', 'on time', and 'too late'. If DFF1 outputs high and DFF2 outputs low, it is the state of the 'on time'. If not, the digital state machine will adjust the delay. Figure 4.16 also shows the state diagram of the digital state machine. The drawback of the automatic delay control is it cannot be adopted in high frequency range because the two DFFs need the startup and the hold times.

The omnidirectional wireless power receiver is implemented by integrated circuit and experimented. Figure 4.17 shows the micrograph. We designed three Rectifiers with Current ZCP Prediction for the omnidirectional wireless power receiving. The digital circuit in the micrograph is the automatic delay control.

Figure 4.18 shows the experimental results of the rectifier. The AC input and DC output power of the rectifier are measured. The AC input power is measured by current probe, while the DC output power is measured by resistors. The highest efficiency is 93.6 % when the load is 100 Ohm.

Figure 4.19 shows the energy loss analysis of the rectifier. When the load is 100 Ohm and the input power of the rectifier is 94 mW, the energy loss on the PMOS and the NMOS transistors is 4.1 mW (4 %), the loss on the driving stage is 2 %, and all other control circuits consume 0.26 mW (0.4 %), which include the delay unit, the overlapping generation circuit, and the automatic delay control. Although the control circuit is the most complex part, it consumes little power. It's noted that only the energy loss of the control circuit is not proportional to the power scale, so the rectification efficiency will not change significantly with regards to the power scale.

In Fig. 4.11, we listed several existing designs of low-power high-efficiency rectifiers. To the best of our knowledge, the best rectifying efficiency is around 80.2 % in the present designs [3]. The proposed rectifier with current ZCP prediction has a clear advantage of the power efficiency of 93.6 % in the low frequency range.

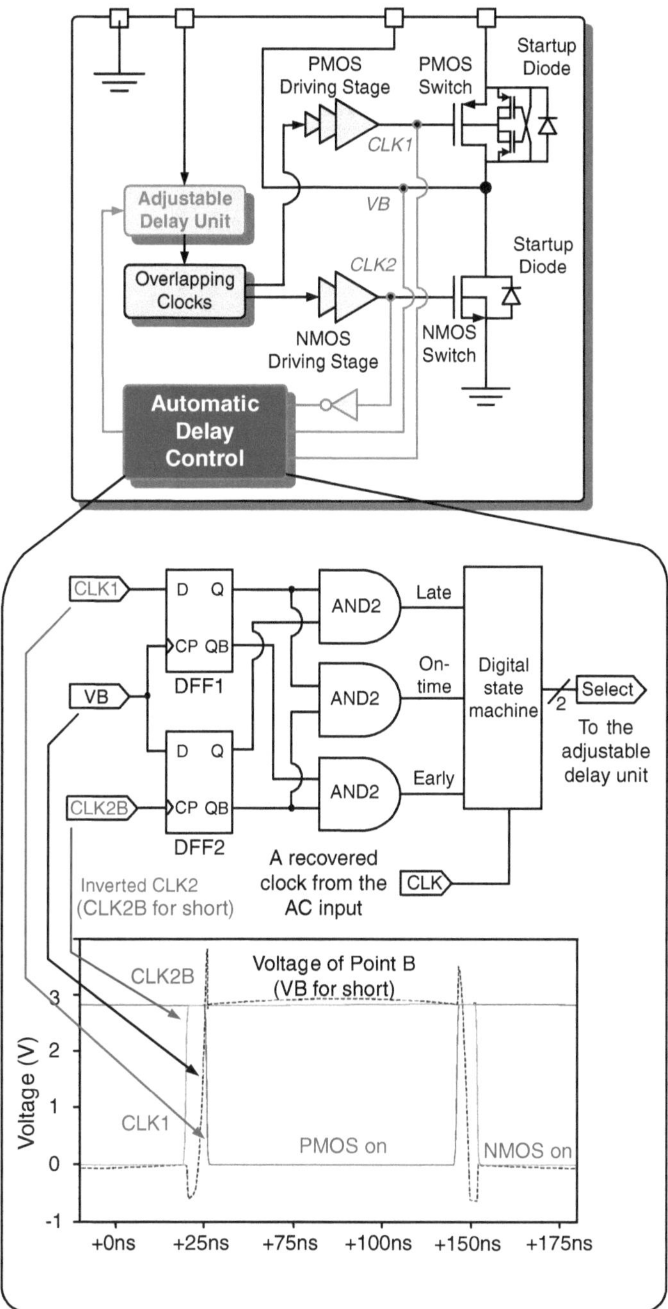

Fig. 4.16 The automatic delay control for the rectifier with current ZCP prediction

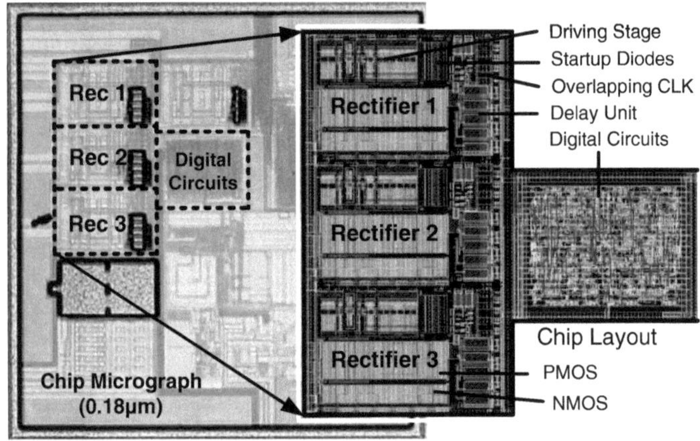

Fig. 4.17 The micrograph of the integrated power receiver with ZCP prediction

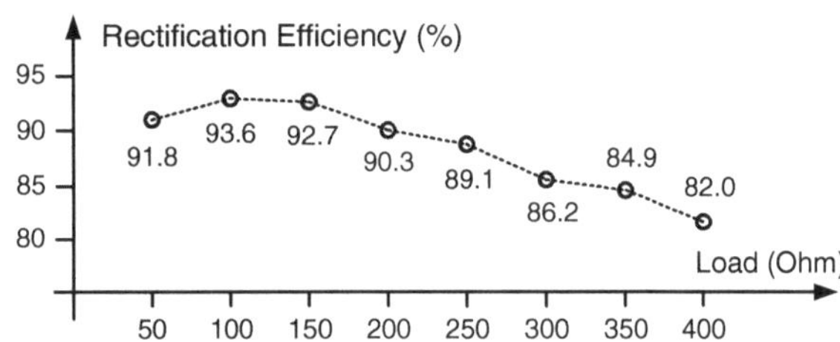

Fig. 4.18 The measured rectification efficiency of the rectifier with ZCP prediction

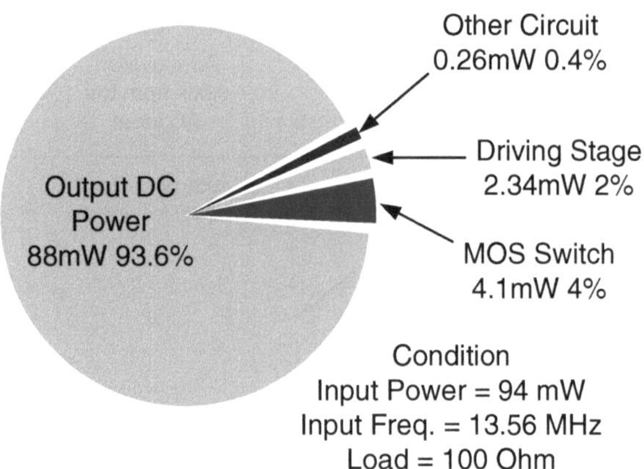

Fig. 4.19 The energy loss analysis of the rectifier

4.2.6 Rectigulator

Some biomedical systems have extremely limited size, like millimeters [6], or even sub-cubic millimeter [8]. Considering the rapid trend of the miniaturization,

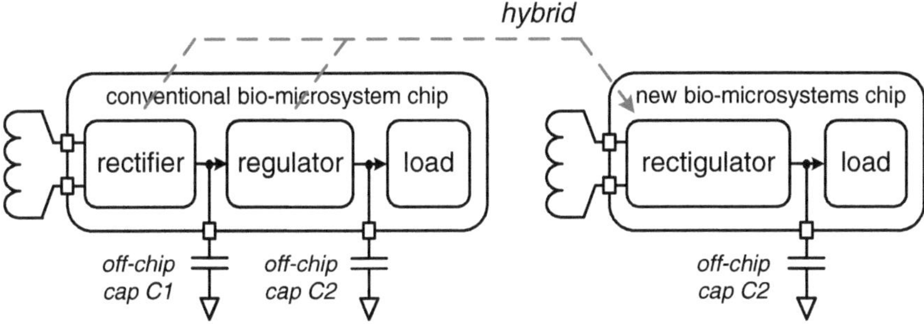

Fig. 4.20 The conventional circuitry and the proposed 'rectigulator'

those devices are expected to be much smaller in the future. In this section, we are going to propose a rectifier for ultra small size biomedical systems.

The implementation of miniature wirelessly powered medical devices is challenging due to the existence of large-size passive components in power receiving circuits. Although capacitor-less power converters exist [15–17], all of them are used in DC–DC applications.

Conventionally, a wireless power receiving circuit consists of an AC–DC converter (rectifier), a DC–DC converter (regulator), and at least two decoupling capacitors C1 and C2 as shown in Fig. 4.20. In most systems, the output decoupling capacitor C1 is too large to be integrated on a microchip.

To save the capacitor area, a new circuitry named the "rectigulator" is proposed by our team [18]. It is showed in Fig. 4.20. The rectigulator is a hybrid circuit of a rectifier and a regulator, which directly converts an AC input to a well regulated DC output. Therefore, the largest capacitor C1 in conventional designs is eliminated.

The proposed rectigulator is showed in Fig. 4.21. It has three inputs (AC1, AC2 and Ref) and two outputs (DC and GND). The inputs AC1 and AC2 are connected to AC power sources like a resonant receiving antenna. The input Ref is connected to a fixed reference voltage like 1.2 V in this design. The reference voltage is used for the regulation of the output DC voltage just as a traditional linear regulator does. The outputs DC and GND provide a well regulated DC power for biomedical microsystems.

As the Fig. 4.21 shows, the rectigulator consists of four parts of components. They are four power transistors, double error amplifiers, double 2–1 analog multiplexers, and a DC voltage divider. Specifically, the power transistors M1–M4 are the main AC–DC paths. The gates of the PMOS M1 and M2 are driven by error amplifiers. The PMOS M1 and M2 work actively and alternatively to allow AC currents flow from the AC1 and AC2 to the DC output. The NMOS M3 and M4 use a cross-gate structure. They switch passively with respect to the AC inputs to allow the DC current flow back from GND to AC1 or AC2. In conventional rectifiers, transistors switch entirely on or off. However, in the proposed

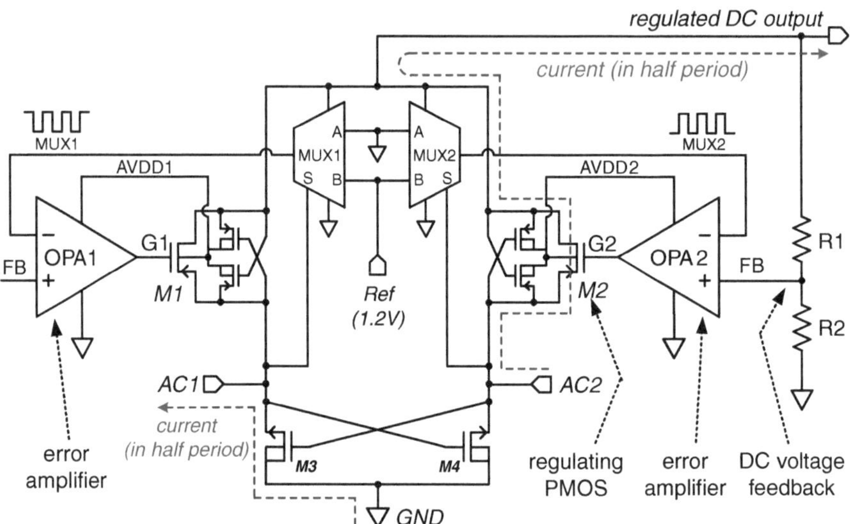

Fig. 4.21 The detail of the proposed rectigulator

rectigulator, the PMOS M1 and M2 act like adjustable resistors. Accordingly, the output DC voltage can be regulated.

The working flow of the rectigulator is described as follows. In each half period, one analog multiplexer outputs ground voltage when the other outputs the Ref voltage. Meanwhile, the DC output voltage is firstly divided by the resistors R1 and R2 and secondly fed back to the error amplifiers. Consequently, one amplifier outputs high voltage to switch off a PMOS, while the other amplifier adjusts the on-resistance of the other PMOS with respect to the DC output voltage and the reference voltage. In other words, the two error amplifiers and the PMOS M1 and M2 regulate the DC output voltage in turns. They ensure the DC output voltage is adjusted to the design value just like a linear regulator does. The designed value can be calculated by Ref (R1 + R2)/R2. It's noted that in order to completely switch off the PMOS M1 and M2, the power source of the error amplifiers are generated by additional transistors using a cross-gate structure as shown in the Fig. 4.21.

Figure 4.22 shows the rectigulator has three operational regions. They are the non-working region, the starting region, and the working region. If the AC input power is not high enough, the DC output may be too weak to activate the circuit, especially the error amplifiers. This is defined as the non-working region. In this region, the AC inputs pass through PMOS without regulation. Consequently, both the DC output voltage and its ripple increase with the growth of the AC inputs. When the AC input power is high enough, the rectigulator enters into the starting region. Due to the start of the regulation, the output ripple decreases with the growth of the AC inputs. If the AC inputs continue to grow, the DC output voltage

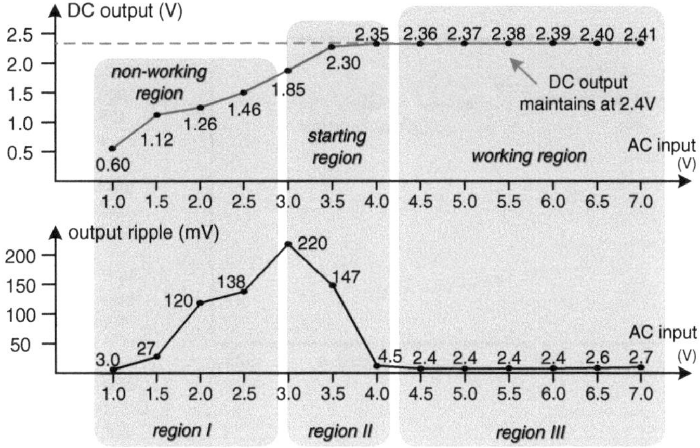

Fig. 4.22 The three operational regions of the rectigulator

will reach and remain at the design value (2.4 V in this design), and the output ripple maintains at a low level. This is the working region.

In the working region, there are some important signals. Figure 4.23a presents waveforms on the OPA2 and Fig. 4.23b indicates waveforms on the PMOS M2. Remarkably, in order to make the voltage G2 reach the voltage of AC2 (to completely cut off the PMOS M2) the OPA2 is powered by the AVDD2, which is generated by a cross-gate structure to dynamically select the higher voltage between the AC2 and the DC output. Finally, circuit verifications on the PCB level is conducted using discrete components as shown in Fig. 4.23c.

The rectigulator saves the area of the decoupling capacitors. According to some typical previous researches, the section areas of the miniature biomedical devices are 4 mm^2 [6], 18 mm^2 [8], and 25 mm^2 [19] respectively. Miniature biomedical devices usually select the SMD package 0402 for off-chip capacitors [19], which occupies at least 0.5 mm^2 (1.0 $\times$ 0.5 mm). As a result, the rectigulator reduces approximately 12 % (0.5 mm^2/4 mm^2) size for the present systems. It's noted that the expected size of biomedical implants is becoming smaller and smaller while the smallest dimension of the off-chip SMD package remains. The existence of off-chip capacitors will be a serious problem.

The main problem of the rectigulator is the power conversion efficiency. Since the AC voltage is directly converted to the designed DC voltage, the magnitude of the AC voltage has to be much higher than the DC voltage. As shown in the Fig. 4.22, when the AC voltage is more than 4.5 V, the DC voltage reaches the designed 2.4 V. Accordingly, the power efficiency of the rectifier is relatively low. Therefore, to use this rectifier, designers have to trade off between the efficiency and the size.

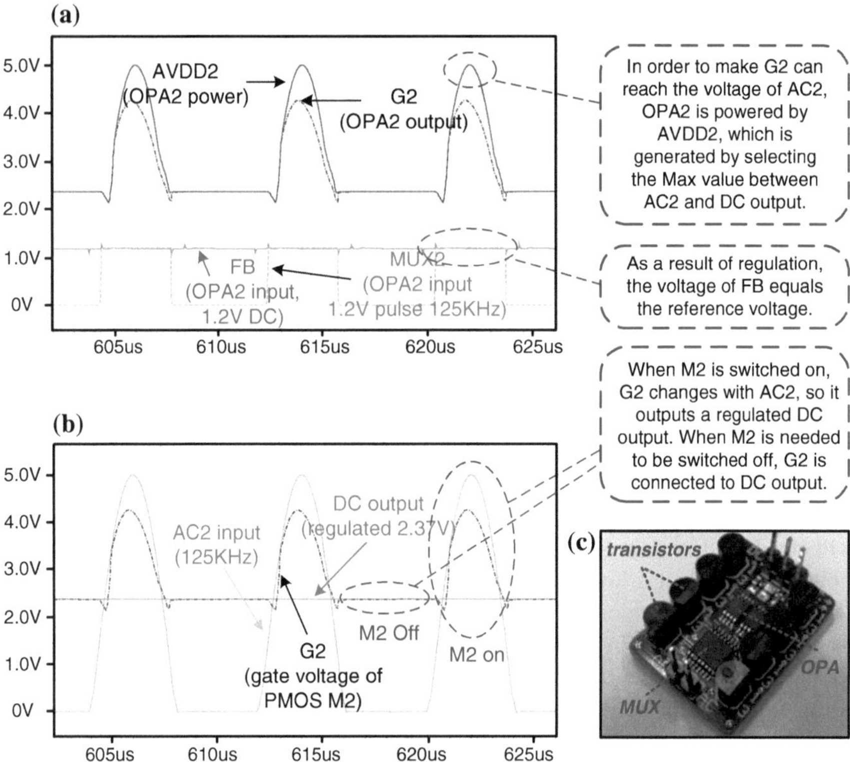

Fig. 4.23 Circuit verification on PCB level and waveforms. **a** Waveforms on OPA2 in working region (AVDD2, MUX2, FB and G2 signals. **b** Waveforms on M2 in working region (AC2, DC, and G2 signals). **c** Circuit verifications with discrete components

4.2.7 Adaptive Rectifier

Unlike battery powered systems that have a stable power source, the wirelessly powered systems suffer from unexpected variations between the transmitter and the receiver. Therefore, to improve the robustness of the power transmission without sacrificing power efficiency, an adaptive rectifier [5] is proposed in GIT's research. The design is showed in Fig. 4.24. The purpose of the design is to allow the wireless power transfer operating over a wider range of transfer distance.

The block diagram shows the reconfigurable active voltage doubler/rectifier. The circuit can be configured into two functions. One is a rectifier when the other is a voltage doubler. The two functions are selected by a comparator. The two inputs of the comparator are a reference voltage and a voltage proportional to the received DC power. If the received power is higher than the reference voltage, the comparator outputs high voltage and configures the circuit to rectifier. If the received power is relative low, the comparator configures the circuits to the

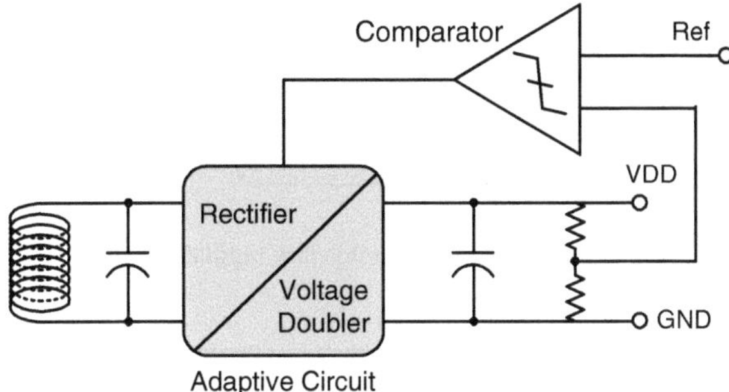

Fig. 4.24 The reconfigurable doubler/rectifier

voltage doubler. In this case, the relative weak wireless power is boosted by the voltage doubler to an acceptable higher voltage for application circuits. In other words, the working range of the wireless power receiver is extended.

4.3 DC–DC Converters

We have introduced the AC–DC converters. In this section, we introduce DC–DC converters. The DC–DC converter is designed to automatically maintain a constant voltage level from another voltage level. In the wireless power transmitter, the DC–DC converter is used to provide DC energy to the inverter. And the in the power receiver, it is used to regulate the unstable output voltage of the rectifier to a stable voltage for the load.

There are two types of DC–DC converters. The first is the linear regulator and the other is the switch-mode DC–DC converter. The two converters are introduced in the following sections. Because the two types of the DC–DC converters are widely used in kinds of electronics besides the wireless power transfer, they are not the key content of this book. We only present several featured designs for the wireless power transfer.

4.3.1 Linear Regulator

The linear regulator is used to maintain a steady voltage. The resistance of the regulator varies with the load and the input voltage. As a result, it acts like a variable resistor. The linear regulator is usually integrated into the wireless power receiver and placed after the rectifier. It outputs a stable voltage for loads.

(a) Capacitor-less Regulator

Although the regulator is very convenient to be integrated into microchips, but the regulator needs off-chip decoupling capacitors. The regular regulator in the power

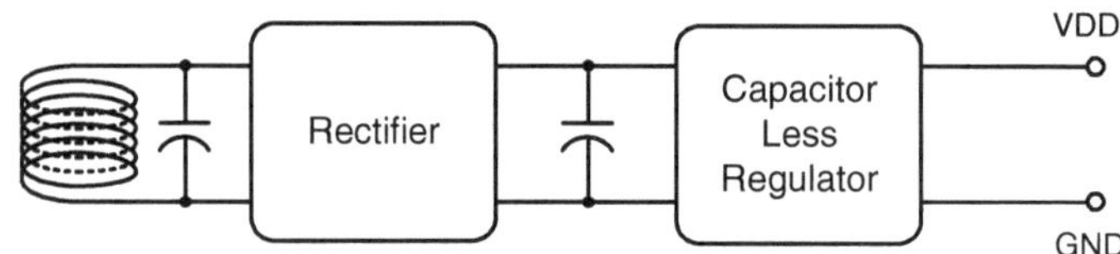

Fig. 4.25 The use of the capacitor-less regulator

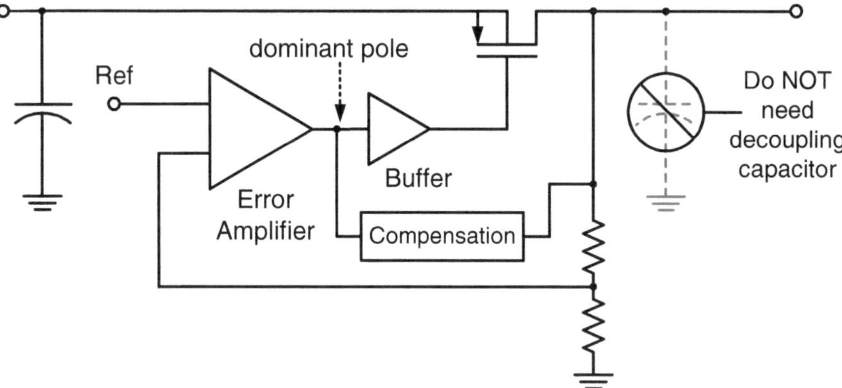

Fig. 4.26 The circuit of the capacitor-less regulator

electronics requires two big decoupling capacitors. They are placed at the input and output terminals respectively. To make the implantable microsystems smaller, researchers tend to reduce the number of off-chip capacitors. As a result, the capacitor-less regulator [15–17] is adopted in the wireless power transfer as Fig. 4.25 shows. In the figure, the capacitor-less regulator needs no decoupling capacitor at the output terminal.

Figure 4.26 shows a typical circuitry of the capacitor-less regulator [16]. The circuit is composed of an error amplifier, a voltage buffer, a PMOS regulation transistor, a resistor divider, and a frequency compensation module.

As the above figure shows, the output voltage is feed back to the error amplifier using the resistor divider. Because the PMOS regulating transistor has very large scale, the buffer is placed after the error amplifier to drive the PMOS transistor. The key is the dominant pole of the circuit is at the output point of the error amplifier, so the feedback loop in the regulator remains stable.

(b) Full-digital regulator

The capacitor-less regulator is achieved by analog circuits. We introduce a fully digital circuit here as the second featured design. Figure 4.27 shows the circuit diagram of the full-digital regulator [20].

The full-digital regulator [20] is composed of a group of parallel PMOS power transistors, a switching controller, and an up/down decider. The PMOS power

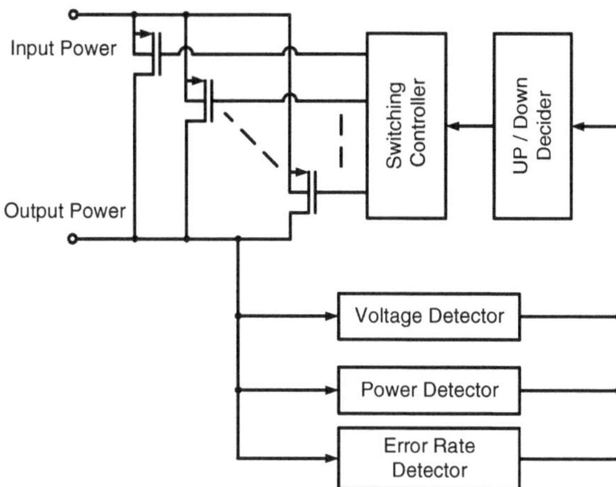

Fig. 4.27 The full-digital regulator

transistors act as resistors. If more PMOS transistors are switched on, the equivalent resistance from the input power terminal to the output terminal becomes smaller, which makes the output power voltage rises. How many PMOS transistors are switched on is determined by the up/down decider. It takes digital information from other feedback control circuits, like the voltage detector, the power detector, and the error rate detector. In the previous research [20], the error rate detector is responsible to evaluate the possibility of circuit operation error due to the decrease of the power supply voltage.

The advantage of the full-digital regulator is that it is very convenient to work with digital circuits. It can be use to provide power supply to digital circuits like digital processors. The power voltage can be easily and digitally adjusted so the digital circuit can work with a dynamic supply voltage. The disadvantage of the full-digital regulator is clear. First, the analysis of the stability of the full-digital regulator is a difficult problem. Unlike the conventional analog regulator has mature methods to analyze the stability of negative feedbacks, the full-digital regulator has no mature method to analyze the stability. As a result, the risk of the design increases. Second, the dynamic output voltage range of the digital regulator is determined by the number of the PMOS transistors. Consequently, if the designer requires a large dynamic output voltage range, there will be too many PMOS transistors.

4.3.2 Switch-Mode DC–DC Converter

Although the linear regulator has a simple structure and it's easy to be integrated into microchips, the power conversion efficiency of the linear regulator is relatively low when the input voltage is much higher than the output voltage. Moreover, the linear regulator can only decrease the voltage but cannot raise the voltage. The switch-mode DC–DC converters convert a DC voltage level to another voltage by temporarily storing the energy in an inductor. In this section, we introduce two featured designs.

(a) Single-inductor Converter

The regular DC–DC converters have one input terminal and one output terminal. It works in the switch-mode. As Fig. 4.31 shows, an off-chip inductor is required to boost or buck the power voltage. However, the biomedical systems usually demand multiple power supply voltages. For example, the digital circuits and random-access memory (RAM) may require a relatively low power supply voltage like 1.2 V. Meanwhile, sensors, analog circuits, and the electrically erasable programmable read-only memories (EEPROM) may need a relatively high power voltage like 2.5 V. As a result, there will be a group of DC–DC converters and corresponding off-chip inductors, which would occupy too much space. To address this problem, many researchers have proposed their single-inductor DC–DC converters [21–23].

Figure 4.28 shows a 1-input 2-outputs single-inductor DC–DC converter. The converter has only one input terminal but two outputs terminals. The main current path of the converter is composed of five power transistors M1–M5 and an off-chip conductor. The inductor and the transistor M1–M3 and M4 constitute the first output channel. Meanwhile, the inductor and the transistor M1–M3 and M5

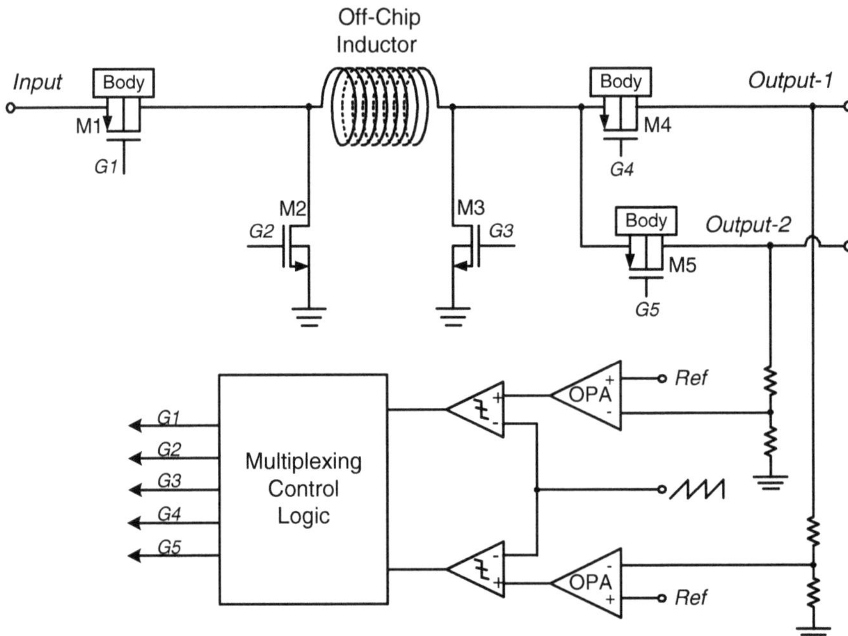

Fig. 4.28 The circuit of multiple-output single-inductor DC–DC converter

constitute the second output channel. Clearly, the off-chip conductor and the M1–M3 are multiplexed by the two channels.

In order to output two different voltages, there are two feedback channels in the above circuits. For each channel, there are an error amplifier, a comparator, and a common control logic circuit. The two output voltages can be feed back by the two resistor dividers respectively. Double error amplifiers are used to track the output voltages. One input of the error amplifier is a reference voltage, and the other input is the feedback voltage. The outputs of the amplifiers are compared with a saw wave and generate PWM waves using two comparators. A multiplexing control logic circuit is employed to use the PWM waves to control the five power transistors in system. The control approach can be based on time division multiplexing for the two channels.

(b) Mode-adaptive Converter

Since the received power level may vary in a large dynamic range, the output voltage of the rectifier varies a lot. As a result, the output voltage of the rectifier may be higher or lower than the voltage required by the load. Accordingly, to use the received energy more efficiently, the DC–DC converter should be able to smartly buck or boost the voltage. Furthermore, in some biomedical applications, a rechargeable battery is employed to temporarily store the received energy. When the received energy degrades, it discharges to supply power for the system. According to these cases, a 2in-2out mode-adaptive converter is proposed by us. One input connects to the rectifier while the other connects to the implantable battery. The mode-adaptive means it smartly changes among three modes. They are the mode of 2-boost, 1-buck 1-boost, and 2-buck. The three modes would be introduced one by one later. Figure 4.29 shows the use of the converter.

In Fig. 4.29, there is a wireless power receiver. It is composed of three rectifiers and the proposed DC–DC converter. The three rectifiers provide an unregulated

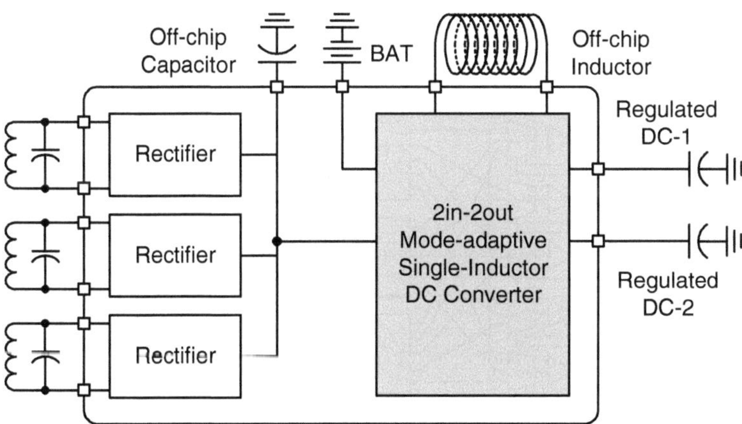

Fig. 4.29 The use of the 2-in 2-out mode-adaptive DC–DC converter

DC voltage. The voltage depends on the received power level. The DC–DC converter is designed to have two inputs and two outputs. One input is tied to the rectifier output while the other connects to a rechargeable battery. The two outputs of the DC–DC converter provide regulated DC power supplies for application circuits. Accordingly, the DC–DC converter is a 2in-2out converter. As the figure shows, the DC–DC converter employs only one off-chip inductor. Figure 4.30 shows the converter with detail.

As the Fig. 4.30 shows, the converter has two input terminals (rectifier and battery) and two output terminals (regulated 1.2 and 2.5 V). To reduce the circuit size, single inductor structure is adopted. Totally six power transistors are used. They are four PMOS transistors and two NMOS transistors. The six transistors constitute totally five possible current paths. They are from Input-1 to Output-1, from Input-1 to Output-2, from Input-2 to Output-1, from Input-2 to Output-2, and from Input-1 to Input-2. The first four paths are all from power sources to loads. The last path is from one power source to another. The first four paths use the common off-chip conductor to reduce the circuit size. The charging and discharging control in the Fig. 4.30 monitors which input source has higher energy level. The input source with the higher power level would be selected to output power for the loads. Additionally, if the Input-1 (rectifier) has higher power, the energy will be charged into the Input-2 (battery) as well as provided to the loads.

The converter has two feedback channels for the two outputs. In each feedback channel, there is a resistor voltage divider followed by a voltage buffer to increase the current driving force. A compensation network is adopted after the buffer to

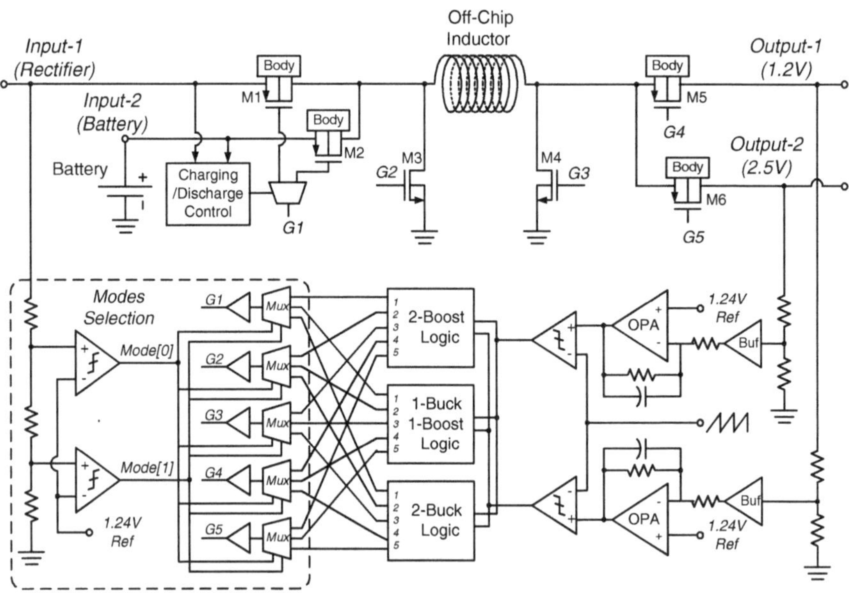

Fig. 4.30 The 2in-2out mode-adaptive single-inductor DC–DC converter

Table 4.1 The three modes in the DC–DC converter

Mode [0]	Mode [1]	Modes
0	0	2-boost
0	1	Impossible
1	0	1-buck 1-boost
1	1	2-buck

Note In order to be adaptive to the wireless power, the mode of the converter automatically changes with regard to the input DC voltage. The Mode [0] and [1] are donated in the Fig. 4.30

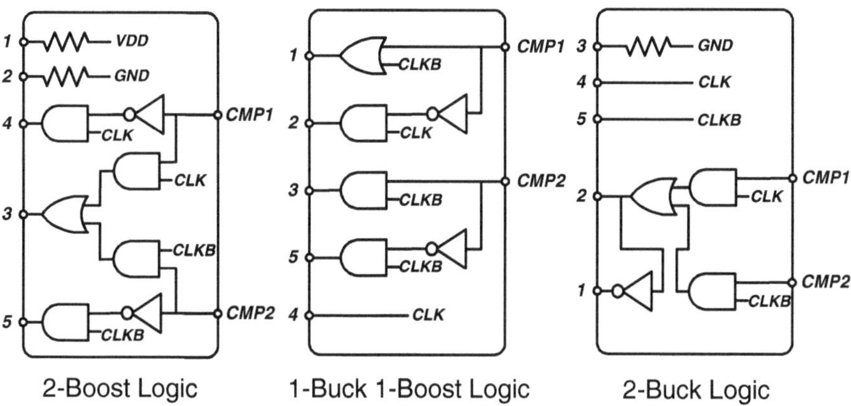

Fig. 4.31 The logic circuits for the three modes in the DC–DC converter

provide enough gain and timing compensation for the feedback channel to ensure the feedback stability. The gain and the compensation are fulfilled by an OPA. The output of the OPA is sent to an input terminal of a comparator. The other input terminal of the comparator is a zigzag wave at a fixed operating frequency. Accordingly, the comparator outputs PWM waves to digital control logic blocks.

Due to the voltage of the Input-1 and Input-2 may be higher or lower than the voltages required by the load, there are three modes in the converter. Table 4.1 shows the three modes.

They are the mode of “2-boost”, “1-buck 1-boost”, and “2-buck”. There is a mode selection module in the converter. Two comparators are adopted to classify the rectifier output into three intervals. They are the intervals of lower than 1.2 V, between 1.2 and 2.5 V, and higher than 2.5 V. The outputs of the comparators are sent to five digital multiplexers to fulfill the mode selection.

Because there are totally three modes, there are three corresponding digital logic blocks in the converter. Their inputs are the PWM waves from the comparators, while their outputs are selected by the five digital multiplexers and used to drive the six switch-mode power transistors. Figure 4.31 shows the three logic circuits in the Fig. 4.30.

The three digital logic circuits all use the operating principles of the time division multiplexing for the two output channels, so the single inductor is multiplexed for two output channels. This is our designed mode-adaptive 2in-2out single-inductor DC–DC converter. It automatically selects the input power source from the rectifier and the battery to provide two regulated and different voltages for the loads. Since there are three modes in the converter, the input voltage range is extended from 0.7 to 3.3 V.

4.4 DC–AC Converters

A DC–AC converter or inverter is an electrical power converter that changes DC energy to AC energy. In the wireless power transfer, the inverter is widely adopted at the transmitting side to convert the DC energy from battery or adapter to alternating power at a specific operating frequency. Typically in the biomedical applications, the input DC voltage of the inverter is in range from 3 to 36 V. And the output frequency of the AC power is usually in the frequency range from 100 K to 50 MHz. The most important features of the inverter are the power efficiency and the operating frequency. Since the power and voltage of the inverters in the power transmitter are relative large, the inverters are typically fulfilled by power MOSFETs and other discrete components on print circuit board (PCB) but not integrated circuit on a microchip. Furthermore, the design of the inverter is relatively mature. We only introduce two basic inverters that are frequently used.

(a) Class D inverter

In Class D inverter or amplifier, all power devices are operated as binary switches. Usually, power MOSFET will be adopted as the switching device. They are either fully switched on or switched off. Figure 4.32 shows a typical Class D power inverter.

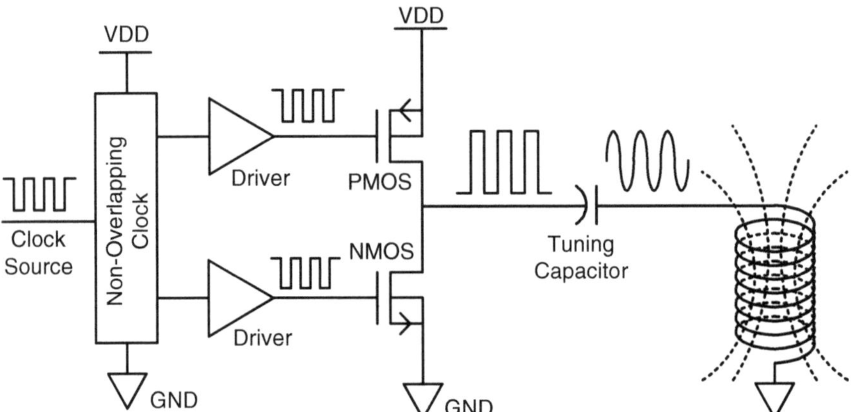

Fig. 4.32 The class D power inverter

As the above figure shows, the Class D inverter consists of a non-overlapping clock generator, two drivers, a PMOS and a NMOS transistor. The clock generator produces two non-overlapping digital clocks. They are amplified by the two drivers and sent to the gate of the PMOS and NMOS transistors. Sometimes, the PMOS transistor is replaced by another NMOS transistor for lower on resistance and higher frequency. The two transistors consists a NOT gate. Accordingly, the inverter outputs a square wave. Since the tuning capacitor and the coil resonate at the designed frequency, the square wave is filtered to a sine wave in the coil.

Figure 4.33 shows two Class D power inverters designed by our team. They are designed to wirelessly power the endoscopic capsule in human body in two different projects. The first inverter has relatively high-power (50 W) and high-frequency (8 MHz) output. It uses large size heat sinks. The second inverter on the right has relatively low-power (30 W) and low-frequency (1 MHz) output. In order to use a smaller heat sink, a cooling fan is employed. The both inverters use the city power adapter as the DC source.

The Class D inverter has two advantages. Firstly, the design of the Class D inverter does not rely heavily on the specific operating frequency. For example, an inverter designed for 13.56 MHz can be also used for lower frequencies, even like 125 kHz. It's only need to adjust the clock input, the tuning capacitor, and the coil. This advantage is very useful for designers because sometimes the frequency would be optimized at the last moment of the project, however, the designing of the inverter cannot wait. The second advantage of the Class D inverter is the power conversion efficiency. The theoretical power efficacy of Class D inverter is 100 %, and in real circuit, the efficiency is usually in range from 70 to 90 %.

The disadvantage of the Class D inverter is it has relative low frequency. Consequently, it is rarely used in high frequency (like over 13.56 MHz) applications. It's noted that most biomedical applications adopt relatively low frequency, like below 1 MHz for less energy loss in human tissue.

Fig. 4.33 Two Class-D power inverters designed by our team for wireless power transfer. **a** Designed high power (50 W) class-D inverter. **b** Designed medium power (25 W) class-D inverter

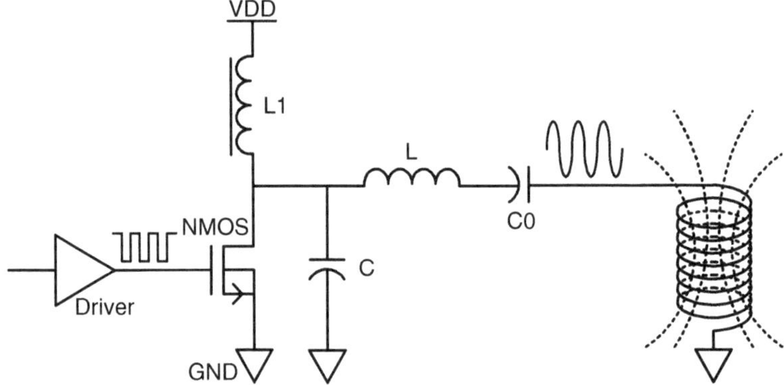

Fig. 4.34 The class E power inverter

(b) Class E inverter

Besides the Class D inverter, Class E inverter [24, 25] is also frequently used in wireless power transfer. The Class E inverter is highly efficient switching power inverter, typically used in high frequencies that the switching time becomes comparable to the duty time at the operating frequency. Figure 4.34 shows a Class E inverter. It consists of a driving circuit, a power MOSFET, a choke inductor, a capacitor C, an inductor L, a tuning capacitor C_0, and a coil. The NMOS transistor is switched on periodically and the energy is transmitted to the resonant tank formed by the tuning capacitor and the coil.

The main advantage of the Class E power inverter is it can be used in the high frequency applications. The most wireless power transmitters working at operating frequencies higher than 13.56 MHz are Class E power inverters. The problem of the Class E inverter is that the values of the adopted components are highly depended on the operating frequency. Once the operating frequency is slightly changed, the inductor and the capacitor in the circuit are needed to be adjusted. In some designs, designers need to find the best operating frequency by experiments. As a result, the Class E power inverter becomes a nightmare because the components in the Class E inverter have to be adjusted again and again.

4.5 Summing Up

In this chapter, we have presented the techniques of the power converter in wireless power transfer for biomedical microsystems. First, a circuit model is given at the beginning of this chapter. The circuit model involved three types of power converters in the wireless power transfer. They are the (1) AC–DC converters or rectifiers, (2) the DC–DC converters or regulators, and (3) the DC–AC converters or inverters.

The AC–DC converters or rectifiers have become the absolute focus of this chapter. It is because the low-power low-voltage rectifiers are not frequently used in common electronics. However, the performance of the rectifier is extremely important for the wireless power transfer. Since the recovered energy might be very low and the operating frequency might be much higher than 50 to 60 Hz, the regular power rectifiers using diodes are solely adopted in wireless power transfer. Instead, researches have proposed many types of new rectifiers. By using cross-gate structure, the self-synchronous rectifier [6] has very simple structure and very high operating frequency, like over GHz. However, it has relative low efficiency, like 65 % [6]. To improve the power conversion efficiency, the comparator-based rectifier [3, 9, 10], the rectifier with ZCP prediction proposed by our group [14], and the full-NMOS rectifier proposed by our group are introduced respectively. Their power conversion efficiencies can easily reach 80 % [3] or even 90 % [14]. For some biomedical devices, smaller size is more important than higher efficiency. Accordingly the rectigulator [18] has been proposed by our team. It combines the rectifier and the regulator together to reduce the use of off-chip decoupling capacitors. For those applications requiring omnidirectional wireless power transfer, the parallel rectifiers [13] proposed by us can be adopted. It picks up energy from multiple coils in different directions.

Besides the AC–DC converters, we have introduced DC–DC converters and DC–AC converters in this chapter. Since they are very widely used in kinds of electrical appliances and their designs are relatively mature, they are not the key point of this chapter. We have briefly introduced several featured DC–DC converters including the capacitor-less regulator [15–17], the full-digital regulator [20], the single-inductor switch-mode DC–DC converter [21–23], and the mode-adaptive DC–DC converter proposed by us. As to the DC–AC converters, we have briefly illustrated the Class D and E inverters, which are frequently used in wireless power transmitters.

References

1. Nakamoto, H., Yamazaki, D., Yamamoto, T., et al., (2006). A passive UHF RFID tag LSI with 36.6 % efficiency CMOS-only rectifier and current-mode demodulator in 0.35 μm FeRAM technology. *ISSCC* (pp. 1201–1210).
2. Yoo, J., Yan, L., Lee, S., et al. (2010). A 5.2 mW self-configured wearable body sensor network controller and a 12 uW 54.9% efficiency wirelessly powered sensor for continuous health monitoring system. *IEEE Journal of Solid-State Circuits, 45*(1), 178–188.
3. Lee, S. B., Lee, H.-M., Kiani, M., et al. (2010). An inductively powered scalable 32-channel wireless neural recording system-on-a-chip for neuroscience applications. *IEEE Transactions on Biomedical Circuits and Systems, 4*(6), 360–371.
4. Ghovanloo, M., & Najafi, K. (2004). Fully integrated wideband high-current rectifiers for inductively powered devices. *IEEE Journal of Solid-State Circuits, 39*(11), 1976–1984.
5. Lee, H.-M. & Ghovanloo, M. (2012). An adaptive reconfigurable active voltage doubler/rectifier for extended-range inductive power transmission. *ISSCC* (pp. 286–288).

6. O'Driscoll, S., Poon A. & Meng, T.-H. (2009). A mm-sized implantable power receiver with adaptive link compensation. *ISSCC* (pp. 294–295).
7. Chiu, H.-W., Lin, M.-L., Lin, C.-W., et al. (2010). Pain control on demand based on pulsed radio-frequency stimulation of the dorsal root ganglion using a batteryless implantable CMOS SoC. *IEEE Transactions on Biomedical Circuits and Systems, 4*(6), 350–359.
8. Chow, E.Y., Chakraborty, S., Chappell, W.J., et al. (2010). Mixed-signal integrated circuits for self-contained sub-cubic millimeter biomedical implants. *ISSCC* (pp. 236–237).
9. Sun, Y., Jeong, C., Han, S., et al. (2011). A high speed comparator based active rectifier for wireless power transfer systems. *MTT-S* (pp. 1–2).
10. Sun, Y., Lee, I., Jeong, C., et al. (2011). A comparator based active rectifier for vibration energy harvesting systems. *ICACT* (pp. 1404–1408).
11. Thidé, B. (2004). *Electromagnetic field theory*. Uppsala: Upsilon Books.
12. Lenaerts, B., & Puers, R. (2007). An inductive power link for a wireless endoscope. *Biosensors & Bioelectronics, 22*(7), 1390–1395.
13. Sun, T. J., Xie, X., Li, G. L., et al. (2012). Integrated omnidirectional wireless power receiving circuit for wireless endoscopy. *Electronics Letters, 48*(15), 907–908.
14. Sun, T., Xie, X., Li, G., et al. (2011). An omnidirectional wireless power receiving IC with 93.6% efficiency CMOS rectifier and skipping booster for implantable bio-microsystems. *A-SSCC* (pp. 185–188).
15. Chen, J.-J., Lin, M.-S., Lin, H.-C., et al. (2008). Sub-1 V capacitor-free low-power-consumption LDO with digital controlled loop. *APCCAS* (pp. 526–529).
16. Kim, Y.-I., & Lee, S.-S. (2012). Fast transient capacitor-less LDO regulator using low-power output voltage detector. *Electronics Letters, 48*(3), 175–177.
17. Chong, S.S. & Chan, P.K. (2012). A 0.9 uA quiescent current output-capacitorless LDO regulator with adaptive power transistors in 65 nm CMOS. IEEE Transactions on Circuits and Systems I.
18. Sun, T. J., Xie, X., Li, G. L., et al. (2012). Rectigulator: A hybrid of rectifiers and regulators for miniature wirelessly powered bio-microsystems. *Electronics Letters, 48*(19), 1181–1182.
19. Harrison, R. R., Watkins, P. T., Kier, R. J., et al. (2007). A low-power integrated circuit for a wireless 100-electrode neural recording system. *IEEE Journal of Solid-State Circuits, 42*(1), 123–133.
20. Hirairi, K., Okuma, Y., Fuketa, H., et al. (2012). 13% power reduction in 16b integer unit in 40 nm CMOS by adaptive power supply voltage control with parity-based error prediction and detection (PEPD) and fully integrated digital LDO. *ISSCC* (pp. 486–488).
21. Chae, C.-S., Le, H.-P., Lee, K.-C., et al. (2009). A single-inductor step-up DC–DC switching converter with bipolar outputs for active matrix OLED mobile display panels. *IEEE Journal of Solid-State Circuits, 44*(2), 509–524.
22. Huang, M.-H., Chen, K.-H. & Wei, W.-H. (2008). Single-inductor dual-output DC–DC converters with high light-load efficiency and minimized cross-regulation for portable devices. *VLSI* (pp. 132–133).
23. Chang, W.-H., Wang, J.-H., & Tsai, C.-H. (2010). A peak-current controlled single-inductor dual-output DC–DC buck converter with a time-multiplexing scheme. *VLSI-DAT* (pp. 331–334).
24. Lee, R., Ibrahim, T.S., Baertlein, B.A., et al. (2001). RF coil modeling and analysis in high field MRI: lessons learned. *Antennas and Propagation Society International Symposium* (pp. 362–365).
25. Acar, M., Annema, A.-J. & Nauta, B. (2007). Analytical design equations for class-E power amplifiers. *IEEE Transactions on Circuits and Systems I, 54*(12), 2706–2717.

Chapter 5
Wireless Power Management

Abstract The circuits in wireless power transfer can be classified to power converters and power management circuits. We present the power management in this chapter. The power management is actually everywhere in the wireless power transfer, especially for those relatively complex systems. In this chapter, we firstly build a circuit model of the power management in the wireless power transfer for biomedical applications. Secondly, we present several detailed designs, including the automatic tuning adjustment, the power adjustment, the wireless power watchdog, the wireless power switch, and the recharging management circuit. These designs are to improve the transfer efficiency and the system reliability, or fulfill other purposes. By using them, the wireless power transfer becomes smarter.

5.1 Introduction

The power management is a general concept in circuit designs. In regular electrical appliances, the power management is responsible to reduce overall energy consumption, prolong battery life for portable devices, and take balance between low power consumption and high performance.

The wireless power transfer is more eager to the power management than regular electrical appliances. It's because the wireless power transfer faces unique and difficult problems as follows:

(a) Problem A: In wireless power transfer, the transfer efficiency depends on the accuracy of adopted tuning components. As introduced before, tuning coils and capacitors are required in the circuits. As a result, their accuracy affects the performance of the circuits. For example, the capacitance precision in the mass production is usually around 5–15 %, which may cause detuning in system. Special circuit techniques are needed to address this problem.

T. Sun et al., *Wireless Power Transfer for Medical Microsystems*,
DOI: 10.1007/978-1-4614-7702-0_5,

(b) Problem B: The power efficiency and the received power level is a function of the transfer distance and the relative orientation between the transmitter and the receiver. The wireless power transfer system is actually a dynamic system. Both the transfer distance and the orientation may vary during the transfer. Accordingly, how to keep the receiver having a stable recovered power is a challenging issue.
(c) Problem C: The wireless power supply and demand at the receiving side is not always matched. In regular electrical device, the supplied power from battery always matches the power consumption of loads. However, in the wireless power receivers, it has to make sure the received energy is at least a little more than it consumes, because the load's power consumption may change with time. Accordingly, energy is wasted and efficiency becomes a problem.

According to these unique problems, researchers have proposed many circuit designs [1–5] for the power management in the wireless power transfer. As you can see, the concept of the power management in the wireless power transfer is not as the same as regular electrical appliances. It emphasizes to promote transfer efficiency and improve the system reliability. To clearly analyze the power management, we firstly show a circuit model of the wireless power management in Fig. 5.1.

In this model, the up side part is the power transmitting circuit, and the down side part is the receiving circuit. In biomedical applications, the transmitting side is commonly placed outside of the human body. It's composed of DC energy source, DC–DC converters [6, 7], DC–AC converters [7, 8], tuning circuits [9–11], power management circuits, and power transmitting antennas [12, 13]. The receiving side

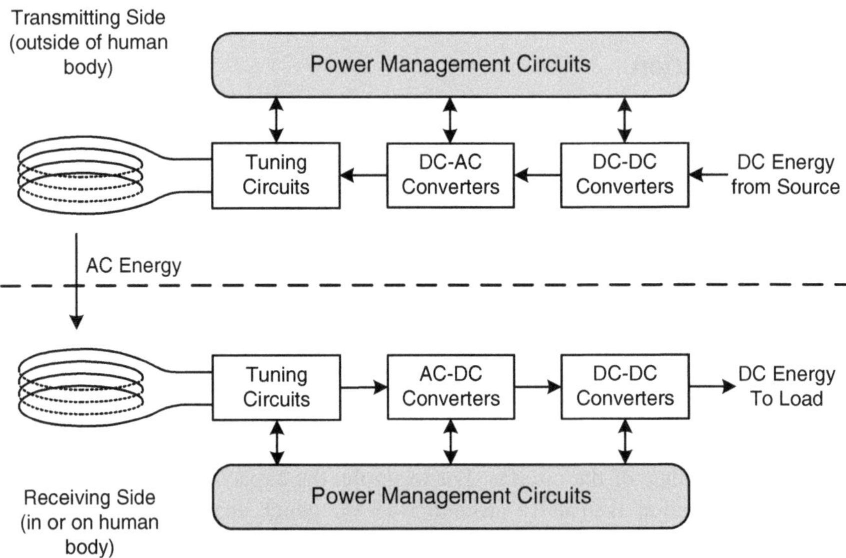

Fig. 5.1 The circuit model of wireless power management

is commonly on or in the human body. It's constitute of power receiving antennas [12, 13], tuning circuits [9–11], AC–DC converters [14, 15], DC–DC converters [6, 16], power management circuits, and loads. The transmission is wirelessly achieved by using couplings between the antennas in space. According to the figure, the power management can be adopted in both sides. It could have connections with all other circuit modules. For example, the tuning circuits can be smartly controlled by the power management circuit to prevent the detuning [9–11]. The AC–DC and DC–DC converters can be managed to adapt to the received power level [1]. Moreover, the power management circuits can be used for unique purposes, like to promote the system reliability, indicate the received power level, reset the whole system in the worst case, and so on. As a result, the power management in the wireless power transfer is a general concept. In this chapter, we introduce the state-of-the-art designs. By using these designs, the wireless power transfer becomes smarter, stronger, and more reliable.

5.2 Tuning Adjustments

All wireless power transfers use resonant circuits to emit and pick up power in space. Tuning circuit is needed to fulfill this task. The simplest tuning circuit is the LC resonant circuits. However, sometimes the operating frequency is needed to be change dynamically, the tuning capacitors cannot be fabricated precisely, and unexpected magnetic or metal objects may change the magnetic field. All these factors affect the resonance status. In other words, the tuning circuit and the wireless power link are sensitive to these factors. Accordingly, some power management techniques [9–11, 17, 18] have proposed to overcome the sensitivities and keep antenna tuning perfectly. These techniques are called the tuning adjustments. They include the capacitor calibration at the transmitting side [10], the capacitor calibration at the receiving side [11], the capacitor calibration at the both sides [9], and finally the frequency tracking [17, 18]. We introduce these techniques one by one as follows.

5.2.1 Capacitor Calibration at the Transmitting Side

If the tuning capacitor at the transmitting side can be dynamically calibrated, the transmitting antenna could resonate better. The technique of the capacitor calibration is needed for the following reasons:

1. Reason A: When the transmitter and the receiver are very close, the wireless power transfer link may be under the status of over-coupling [19]. Accordingly, the phenomenon of the frequency splitting [19] would happen. The power efficiency can be improved by adjusting the capacitance value of tuning capacitor.

2. Reason B: Circuits under development in Lab can be fabricated very precisely. However, to maintain a balance between low cost and high performance, the component precision in the mass production is usually only 5–15 %. Consequently, to make sure all circuits resonate in the best state, the calibration of the tuning circuit in the transmitting circuit is necessary.
3. Reason C: Unexpected magnetic or metal objects may disturb the magnetic field generated by the wireless power transmitter. The patient's movement may also change the transfer distance and the relative orientation. The calibration of the tuning circuit could eliminate the sensitivity to the unexpected interference.
4. Reason D: In order to make patient more comfortable, the transmitting antenna in large size may be fabricated to be soft or flexible. It can slightly vary its shape accordingly to the patient's movement. Consequently, the inductance of the antenna varies. Therefore, the calibration is required to ensure the resonance.

We have understood the reason why the calibration of the tuning circuit at the transmitting side is necessary. However, which component are we going to calibrate? Is the inductor or the capacitor? The inductor in the tuning circuit is actually a coil acting as an antenna to pick up magnetic energy in space, while the capacitor is always a component on PCB or on chip. Compared to the inductor, the capacitor is much more convenient to be adjusted. We only need to parallel more or less capacitors to adjust the capacitance. Therefore, the most calibrations are based on the adjustment of the tuning capacitor but not the inductor. Figure 5.2 shows the circuit model of the capacitor calibration at the transmitting side.

As shown in Fig. 5.2, there is only single tuning capacitor at receiving side, but there are two parallel tuning capacitors at transmitting side. One is a constant value capacitor acting as the basic tuning capacitor, and the other is an adjustable capacitor for the capacitor calibration. The capacitors can be placed on PCB or on microchip. The adjustable capacitor is usually achieved by a capacitor array and there are two ways to constitute the capacitor array as Fig. 5.3 shows. The first way uses parallel capacitors with the same capacitance [11]. Capacitors are connected together using switches. The switches can be made of power transistors. If

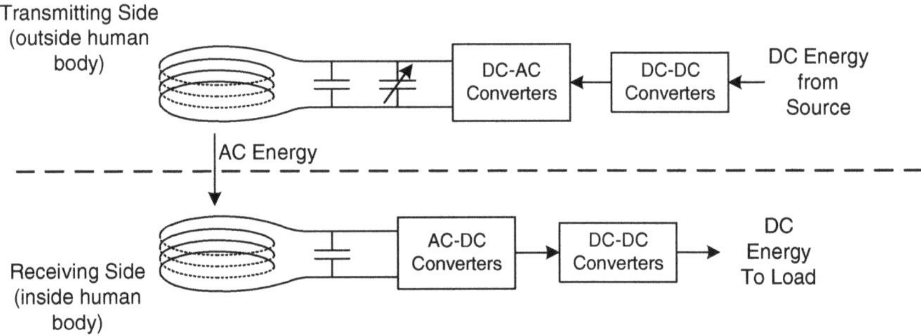

Fig. 5.2 The circuit model of capacitor calibration at the transmitting side

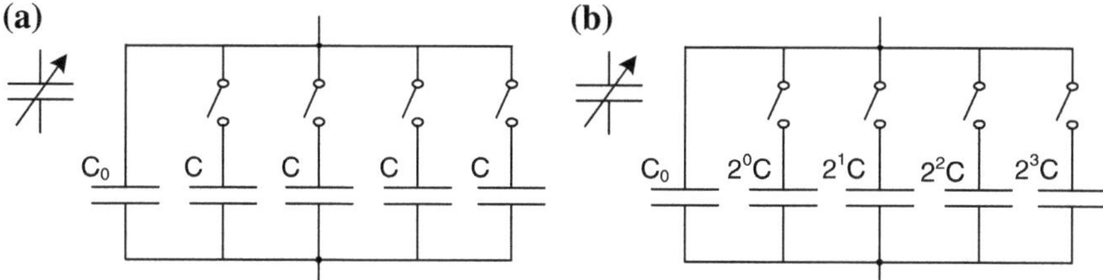

Fig. 5.3 The adjustable capacitor arrays

all switches are turned on, the equivalent capacitance reaches its maximum value. The second way to constitute the adjustable capacitor is to use parallel capacitors with capacitance in exponential growth [9]. There is no essential difference between the two ways. The first way uses un-weighted capacitors and the second way uses weighted capacitors.

How to calibrate the tuning capacitor? There are two ways to digitally control the adjustable transmitting capacitor. The first way uses a feedback from the receiving side. Its advantage is the calibration is more precise. The second way uses a feedback from the transmitting antenna. Its advantage is it has relatively simple control circuits.

(a) Feedback from the Receiving Side

Figure 5.4 shows the capacitor calibration control using the feedback from the receiving side. The power transmitter is composed of a DC–DC converter, a DC–AC converter, a capacitor controller, and a RF transceiver module. The power

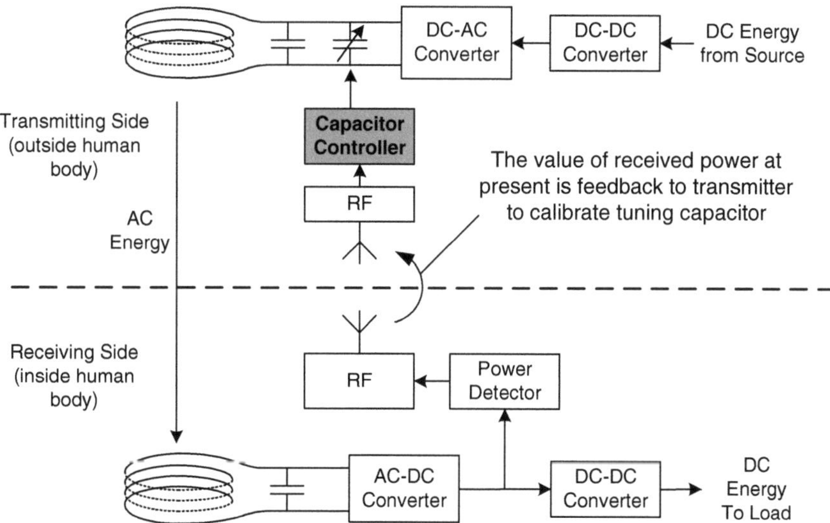

Fig. 5.4 The capacitor calibration control using feedback from the receiving side

receiver is composed of an AC–DC converter, a DC–DC converter, a power detector, and a RF data transceiver. The power detector in the receiver detects the DC output voltage of the AC–DC converter. The voltage information is wirelessly transmitted to the transmitting side by using the RF data transceiver. The capacitor controller adjusts the capacitor according to the voltage information. The ultimate purpose of the calibration is to maximize the DC voltage at the secondary side.

In the method, the adjustment of the capacitor usually needs multiple steps to reach the optimum value. For example, a work [9] proposed by Stanford University requires 18 steps to get the optimum capacitance value. In each step, the calibration control circuit firstly changes the capacitance, secondly waits for new results transmitted from the receiver side, thirdly determines the next value of capacitance, and finally enters into the next step. The capacitance calibration digital control logic can be realized by finite state machine.

The advantage of this method is it has very straight purpose, which is to maximize the received power at the receiving side. By using this method, the ultimate purpose of the improvement of the received power can be assured. However, there are also some disadvantages in the method. The main problem is it requires relatively complex and power consuming circuits, like the power detection and the RF module at the receiving side. Accordingly, this method is not appropriate for those systems requiring very a simple or low-cost design at the receiving side. Another problem of this method is the feedback needs time. As a result, the feedback is usually fulfilled periodically, which makes the system response slowly and less effective.

(b) Feedback from the Transmitting Side

Figure 5.5 shows another capacitor calibration control method. This method only uses the information at the transmitter side. The transmitting side is composed of a DC–DC convert, a DC–AC converter, a power monitor, and a capacitance control module. The AC voltage magnitude on the transmitting antenna is measured by the power monitor. The voltage value reflects the resonance status of the transmitting antenna. The purpose of the capacitor controller is to adjust the capacitance and keep the AC voltage of the transmitting antenna as high as possible. The whole calibration works independently at the transmitting side. By improving the resonance of the transmitting side, it improves the overall transferred power.

The first advantage of the feedback from the transmitter itself is that the design can be very simple and has relatively low cost. Another advantage is the feedback can be very fast, which could keep the transmitter always working at the best condition. Of course, there are disadvantages. Sometimes, the optimization at the primary side doesn't mean the optimization of the whole transfer. For example, if the resonant frequencies of the transmitter and the receiver are slightly different. As a result, the calibration method would probably adjust the tuning capacitor to allow the transmitting antenna working on its resonant frequency. However, at the transmitter's frequency, the secondary side doesn't receive the most energy.

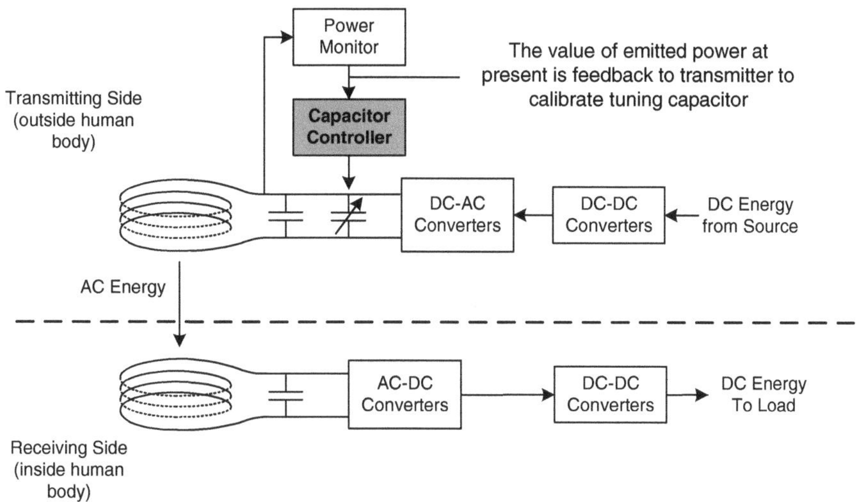

Fig. 5.5 The capacitor calibration control using feedback from the transmitting side

5.2.2 Capacitor Calibration at the Receiving Side

Besides the method to calibrate the tuning capacitor at the transmitting side, capacitor calibration can be also fulfilled at the receiving side [11]. The reasons that the tuning capacitor at the receiving side is required to be calibrated are as same as at the transmitting side. These reasons include the phenomenon of over-coupling [19], the fabrication precision, the unexpected interference from magnetic or metal objects, and so on.

Figure 5.6 shows a simplified circuit model of the capacitor calibration at the receiving side. The receiving side adopts two capacitors. One is a constant value capacitor, and the other is an adjustable capacitor, which is composed of switched capacitor array. Because the size limitation on the receiving side is usually much stricter than the limitation on the transmitting side, all adjustable capacitor array and the switches are designed based on on-chip capacitors and transistors.

Figure 5.7 shows a circuit model of the capacitor calibration at the receiving side. The receiving circuit consists of a DC–DC converter, an AC–DC converter, a power monitor, and a digital capacitor control module. The output DC voltage of the AC–DC converter is monitored by the power monitor. The digital capacitor control module firstly changes the capacitance of the adjustable capacitor, secondly reads the new result from the power monitor, and finally determines the new capacitor value.

A special problem of the capacitor calibration at the receiving side is that the calibration operation may power down the receiver itself just like a self suicide. The calibration operation is actually a serious of tries to find out whether the received power can be higher or not. If the calibration operation adjusts the capacitance in the wrong direction, the received power goes low, which may

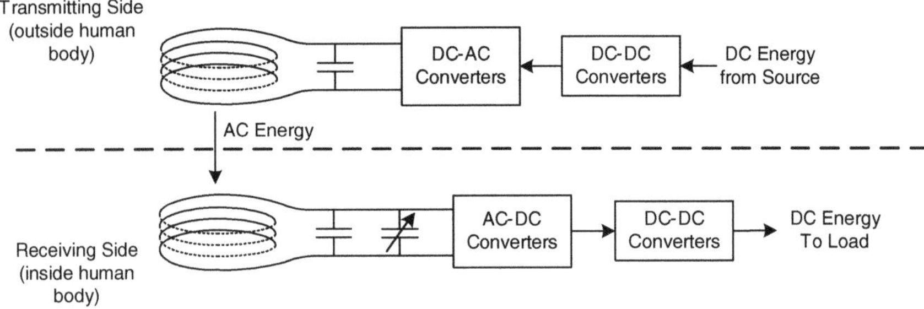

Fig. 5.6 The circuit model of capacitor calibration at receiving side

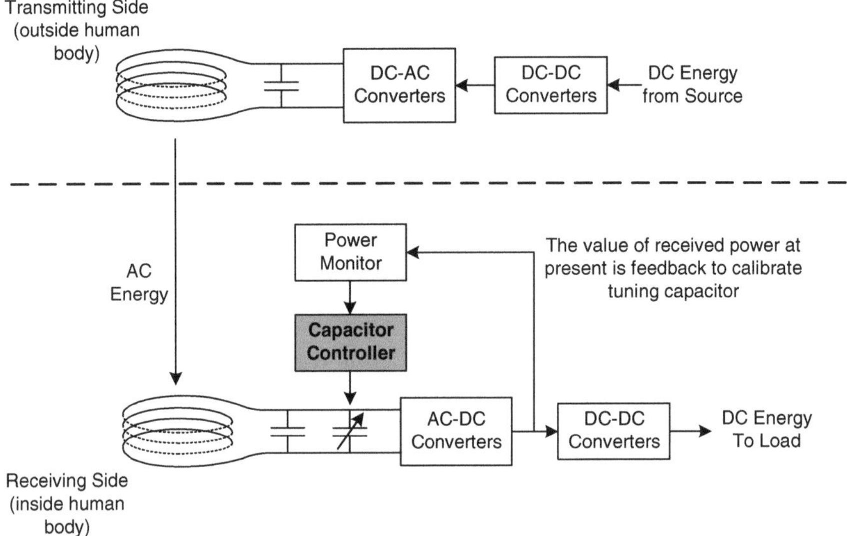

Fig. 5.7 The capacitor calibration at the receiving side

power down the receiver itself. This situation is quite difficult to predict. To our best knowledge, there is still no perfect circuit or algorithm solution for this problem.

We have introduced the methods to calibrate capacitors at the transmitting and receiving side. A natural question is which method gets more performance improvement. Is the calibration at the transmitting side or at the receiving side? Usually speaking, the quality factor of the transmitting side is much higher than the quality factor of the receiving side because the internal resistance of the DC–AC converter is much smaller than the resistance of load. In other words, the resonating circuit at the transmitting side is more sensitive to the detuning problem. Consequently, the calibration at the transmitting side is more necessary.

5.2.3 Capacitor Calibration at the Both Sides

Some work [9] calibrates capacitors at the both transmitting and receiving sides. It makes the circuits and the control algorithm more complex. Figure 5.8 shows the circuit model of the capacitor calibration at the both sides. The transmitter consists of a DC–DC converter, a DC–AC converter, a capacitor control, and a RF transceiver. The receiver consists of a DC–DC converter, an AC–DC converter, a capacitor control, a power monitor, and a RF transceiver. The output DC voltage of the AC–DC converter is detected by the power monitor. The voltage information is sent to two modules. One is the capacitor control module in the receiving side. The other is wirelessly transmitted to the capacitor control module at the transmitting side using the RF transceivers. The two control modules calibrate their adjustable capacitor arrays according to the voltage information.

Because the calibrations at the transmitting and receiving sides can be operated independently, there are actually two degrees of freedom in the system. To ensure the calibration at one side doesn't interfere with the other calibration at the other side, one capacitor will be fixed when the other capacitor is being calibrated. Consequently, the calibration at the both sides has more steps and requires more time. By using the calibration at the both, previous research work [9] reported a power efficiency improvement from −35.8 to −32.1 dB.

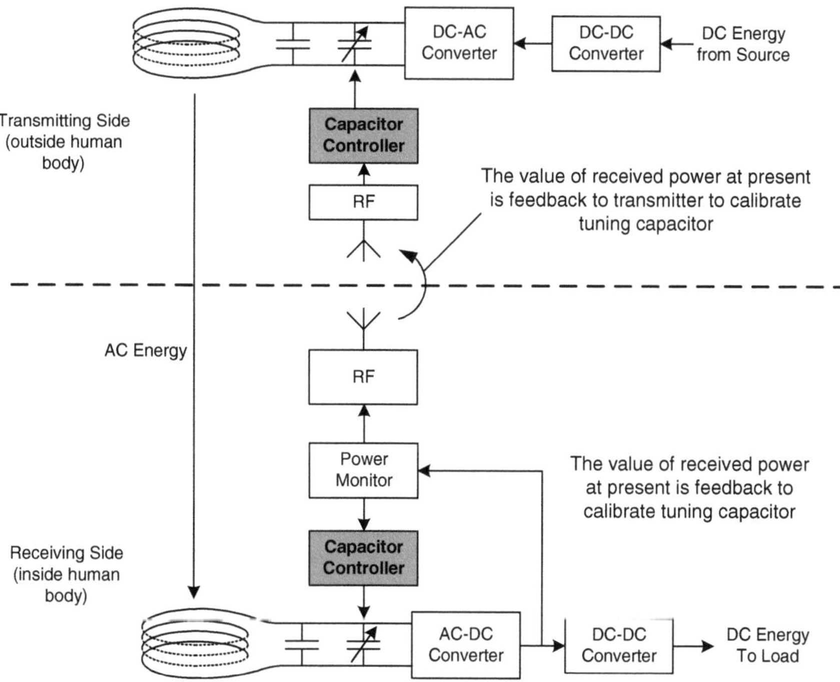

Fig. 5.8 The capacitor calibration at the both sides

5.2.4 Frequency Tracking

In previous section, we have introduced the resonance can be improved by the capacitor calibration. In some researches [17, 18], the frequency tracking is another power management technique for the tuning adjustment. In this technique, the operating frequency is not constant but adjustable. As shown in Fig. 5.9, the transmitting side consists of a DC–DC converter, a DC–AC converter, a power monitor, a frequency control module and a RF transceiver. The operating frequency is determined by the frequency control module. The receiving side consists of an AC–DC converter, a DC–DC converter, a power monitor, and a RF transceiver.

Since the inductors and the capacitors are constant in Fig. 5.9, the only way to optimize the resonance status is the adjustment of the operating frequency. Specifically, there are two methods to determine the operating frequency.

(a) Feedback from the transmitting antenna:

The first method to determine the operating frequency is to use a feedback from the transmitting antenna. By using an analog to digital converter to monitor the voltage and the current on the transmitting antenna, the power monitor detects the emitted power level. A frequency controller reads the result of the power level and

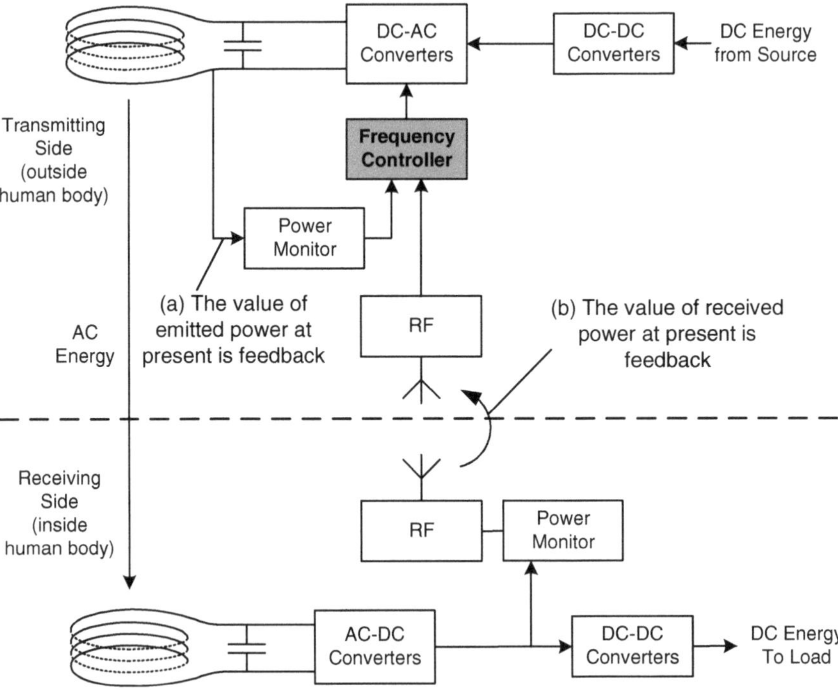

Fig. 5.9 The frequency tracking

adjusts the operating frequency. By adjusting the frequency, the emitted power can be maximized, which means the transmitting antenna has worked in the best resonant status. Since there is no control circuit in the power receiver, this method is suitable for those applications requiring simple receivers.

(b) Feedback from the DC output at the receiving side:

The second method uses a feedback from the DC output voltage of the AC–DC converter at the receiving side. The power monitor at the secondary side would read the voltage information through an analog to digital convert. This voltage information is wirelessly transferred from the receiving side to the transmitting side. A frequency control module reads the voltage information and adjusts the operating frequency. The ultimate purpose of the feedback is to maximize the DC output voltage of the AC-DC converter at the receiving side. The problem of this method is clear. Since this method consumes power at the secondary side, more energy is needed to be transferred.

5.3 Power Regulations

The power regulation is another type of power management in wireless power transfer. In this section, we present the power regulation.

First of all, why the power in wireless power transfer systems is required to be regulated? To answer this question, we have to think how to determine the transmitting power in a regular transfer system. Actually, it can be a very complex question. The reasons may include:

(1) For some applications, typically like the implantable on-demand stimulators [20], transmitters sleep in the most of time. Once the transmitter detects a receiver, it may automatically start to transmit wireless power. Accordingly, there should be a mechanism to detect the existence of the receiver and control the transmitted power. (2) The transfer efficiency and the received power level is a function of the transfer distance and the relative orientation between the transmitter and the receiver. Moreover, both the transfer distance and the orientation may vary in the real applications. So, the wireless power transfer is actually a dynamic progress. How to keep the receiver having a stable received power level is a challenging issue. (3) The power consumption of the receiver may also change. The application circuits at the receiving side may have multiple working statuses, like a low-power mode, a high-performance mode, a suspend mode and so on. Correspondingly, the power consumption varies with modes, which also affects the requirement of the transmitting power.

The above reasons make the determination of the transmitting power quite a difficult problem. To ensure the received power at the secondary is enough but not overmuch, power regulation has been proposed [1, 3]. Overall speaking, existing designs can be summarized in two types. One uses a feedback from the transmitting antenna, and the other feeds back from the receiving side.

5.3.1 Feedback from the Transmitting Antenna

Figure 5.10 shows the power regulation using a feedback from the transmitter.

In Fig. 5.10, the transmitter consists of a DC–DC converter, a DC–AC converter, and a power monitor. The power monitor detects the voltage and the current on the transmitting antenna. By doing so, the output voltage of the DC–DC converter can be adjusted to change the power on the transmitting antenna. How could this circuit help the system design? Actually, it can be used at least for three functions:

(a) The first function is to regulate the transmitting power. Sometimes, designers tend to keep the transmitter emitting at fixed power level. However, the input voltage and the surroundings may change the transmitting power level. As a result, this circuit makes sure that the emitted power level is stable. It is noted that in this function, although the transmitting power is regulated, the received power still varies with the transfer distance and the relative orientation.
(b) The second function is to detect the existence of the receiver. This function is very useful for some applications. For example, the transmitter can be designed to work at low power mode in most of time. The transmitter would change to a high power mode when the voltage on the transmitting antenna varies, which means another resonator appears. The transmitter only works when the power receiver is there.
(c) The third function is to locate the receiver. Why do we need to locate the receiver? Let's give an example. If a patient feels pain at lumbar vertebrae, doctors may implant an electrical stimulator into his lumbar vertebrae. If the patient feels pain, he handles a transmitting near to his waist. However, he does not know the exact location of the implanted simulator in this body. When the transmitter approaches the simulator, the mutual inductance increases. Accordingly, the voltage and the current on the transmitting antenna

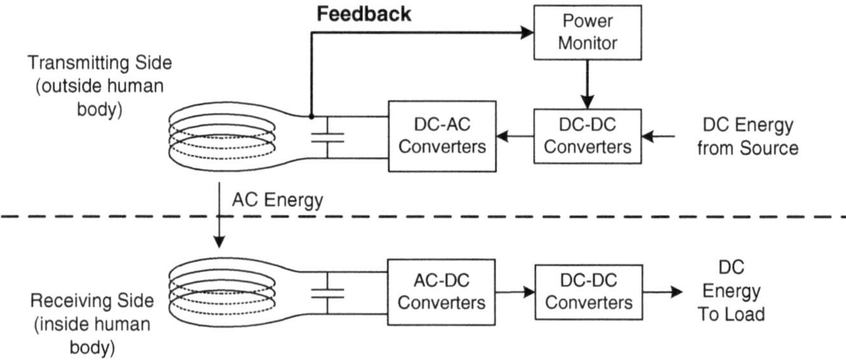

Fig. 5.10 The power regulation using the feedback from the transmitting antenna

change with the mutual inductance. Listening to the sound made by the monitor in the transmitter, the patient can find the best location to transfer energy, just like mine hunting.

5.3.2 Feedback from the Receiving Circuit

As mentioned, there are two types of power regulations. The second type regulates the power using a feedback from the power receiver. Besides the receiver existence detection introduced before, feeding back from the receiver has two more functions. First, it can be used to regulate the received power level, which significantly promotes the system reliability. Second, using the feedback channel, the power receiver could negotiate with the power transmitter about the power level. For instance, when the power receiver plans to enter into the high-performance mode, it could contact with the power transmitter and ask for more power. It makes the system design very flexible.

How to feedback the information from the receiving side? There are two ways.

(a) The First Way: Using the scatter back

As Fig. 5.11 shows, the transmitter is composed of a DC–AC converter, a DC–DC converter and a decoder. The receiver is composed of an AC–DC converter, a power detector, a DC–DC converter, and an adjustable resistor. The power detector firstly monitors the received power level. It secondly adjusts the resistance of the resistor. The change of the resistance affects the power consumption of the receiver. Accordingly, this change is scattered back to the transmitting antenna due to the mutual inductance between transmitting and receiving coils. The decoder at the transmitting side decodes the information and adjusts the transmitting power level.

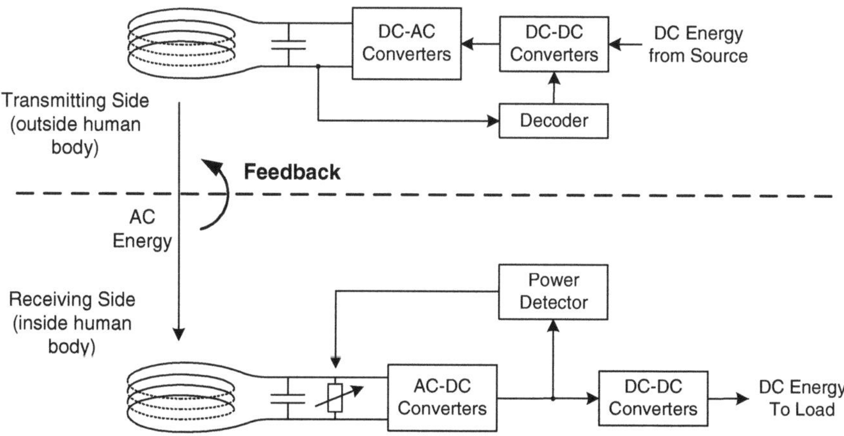

Fig. 5.11 The power regulation using the scatter back from the receiver

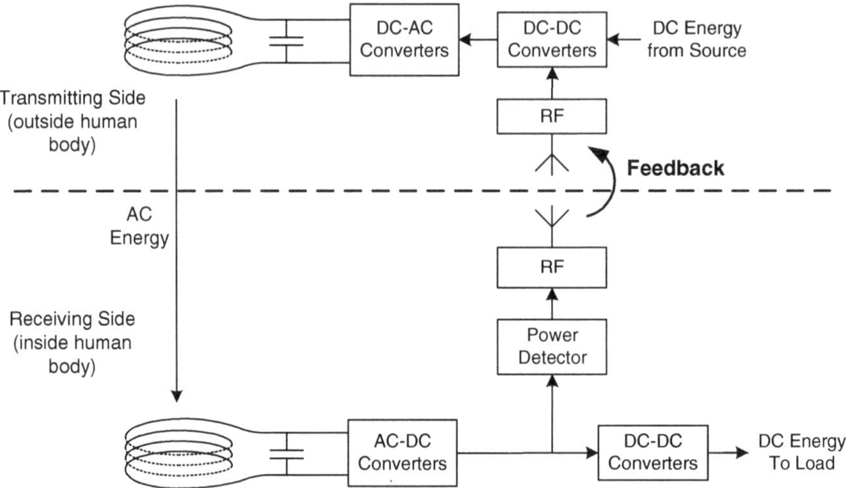

Fig. 5.12 The power regulation using the RF transceivers from the receiver

The advantage of this circuit is that it's very simple and the cost is quite low. The disadvantage is it can be used only in short transfer distance applications because it is quite difficult to scatter back the information to the transmitter if the mutual inductance is too small.

(b) The second way: Using RF transceivers

The second way uses RF transceivers to feedback as Fig. 5.12 shows.

As Fig. 5.12 shows, the transmitter is composed of a DC–AC converter, a DC–DC converter, and a RF transceiver. The receiver is composed of an AC–DC converter, a DC–DC converter, a power detector, and a RF transceiver. The power detector quantizes the received power level and transfers the information back to the transmitter by using the RF transceivers.

The advantage of the circuit is that it can be used in applications requiring middle or long transfer distance. The disadvantage is the RF transceivers consume relative large power, usually measured in milliamperes. Accordingly, the feedback cannot be operated very frequently, which pushes up the overall power consumption of the receiver.

5.4 Wireless Power Watchdog

Many digital microcontrollers use watchdog to ensure the reliability of hardware and software. The simplest watchdog is actually a timer. The software must clear the timer periodically. If the system enters into any unexpected situation, the watchdog wouldn't be cleared in time and the watchdog triggers the reset for the whole system. A complex watchdog may have more functions like fulfilling a procedure before the system reset. The watchdog is very helpful to promote the reliability of the system.

What is a "wireless power watchdog"? Actually, it's a circuit concept proposed by our team. It's a circuit to monitor the received wireless power level, feedback power information to a corresponding power transmitter, and interact with loads. We propose the concept to promote the reliability of wireless power transfer systems. Actually, systems using the wireless power transfer can be very unstable and unexpected. The essential reason of the instability of the wireless power transfer is it hasn't a good negative feedback.

Figure 5.13 shows a regular wired power system. It uses a feedback from the output to ensure the stability of the output power. The output voltage or the current is ensured to be the designed value. Conventional power converters like the LDO [21, 22] and the DC–DC converter [6, 16] all use the feedback technique to keep the output regulated, stable, and capable of responding to input and output changes rapidly. For example, the linear regulator uses resistor voltage divider to track the output voltage.

However, in a wireless power transfer system as illustrated in Fig. 5.14, maybe there is no feedback from the power receiver at all. Even if some systems have feedbacks, the feedbacks depend on the receivers have recovered enough power. If the power is not successfully transferred to the receiver, there is still no feedback. This makes the wireless power transfer somehow out of control. Another reason that the wireless power transfer system is more unstable is that the transfer distance and the relative orientation dynamically change. For example, in the application of

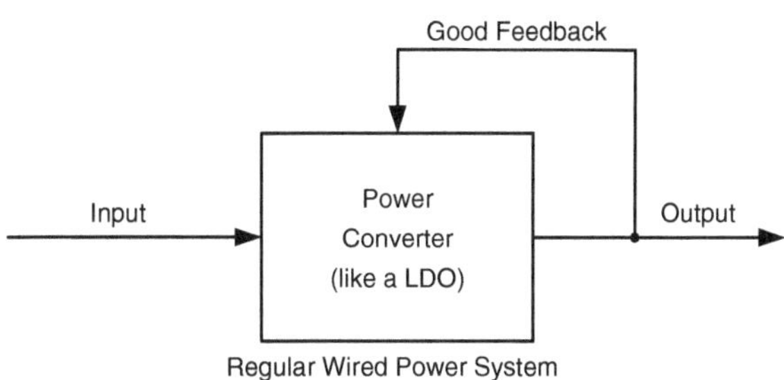

Fig. 5.13 Regular wired power systems with good feedback

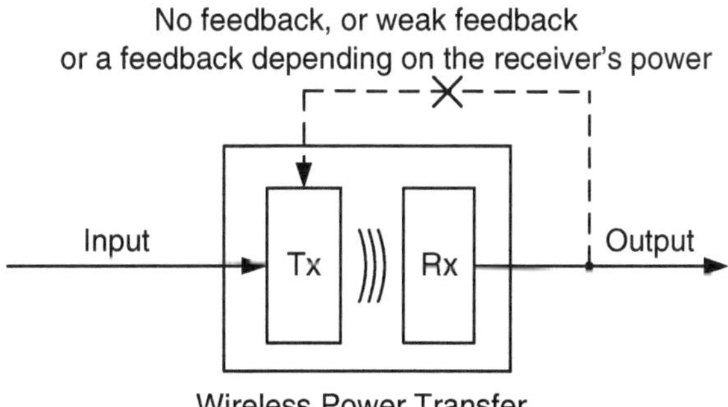

Fig. 5.14 The wireless power transfer with weak feedback

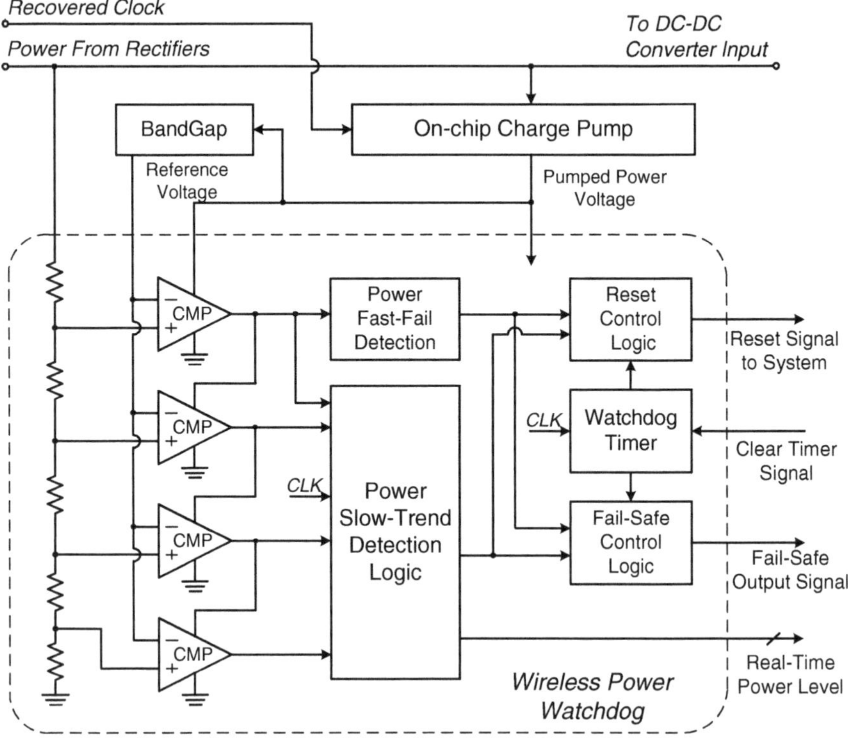

Fig. 5.15 The wireless power watchdog

the capsule endoscopy, the capsule passively moves in the digestive track. In order to improve the system stability and the reliability, we proposed the wireless power watchdog as Fig. 5.15 shows.

As Fig. 5.15 shows, to keep the watchdog the most reliable module in system, a private charge pump is occupied by the watchdog. The input of the charge pump is the output power from rectifiers, and the output of the pump is dedicated for the watchdog. The designed watchdog has three functions. Each function corresponds to a circuit module. (1) First, the watchdog monitors and feedbacks the received power level. As figure shows, four comparators are used to quantize the received power level into five intervals. The module of the power fast-fail detection is responsible to catch voltage ripples that are serious enough to fail the application circuits. The module of the power slow-trend detection is used to monitor the slow change of the power voltage. Application circuits use the power information from the two modules and adjust their power requirement. The power information is also feed back to the transmitter. (2) The second function of the watchdog is that it triggers fail-safe control if the received power level is too low or the application circuits don't clear the watchdog timer periodically. (3) The last function of the watchdog is that it would reset the whole system if the power voltage continuous to fall.

5.5 Wireless Power Switch

Biomedical implants are usually sealed and designed to be waterproofing. The waterproofing design is necessary but it also raises some problems. One of the problems is how to switch on the battery inside of the implants when we hope to startup the system.

There are some conventional methods. For example, designers may use magnetic controlled switch. When a magnet is removed away from an implant, the battery in the implant is switched on and the whole system starts up. The problem of this method is that it can be easily interfered by other unexpected magnetic objects. Another way is to use a RF transceiver in the microsystem. It consumes the energy of the battery and waits the wireless command to startup the system. The problem of this way is the RF transceiver actually consumes a lot of power, like around milliamperes. The battery may have already run out before the system is supposed to be activated.

Figure 5.16 shows our proposed wireless power switch. The most important advantage of the design is its power consumption is nearly zero. Specifically, at the transmitting side, the circuit is composed of a DC–DC converter, a power amplifier, an oscillator, a modulator, and a control circuit. The control circuit sends out a command for switching on the battery. The command is coded and

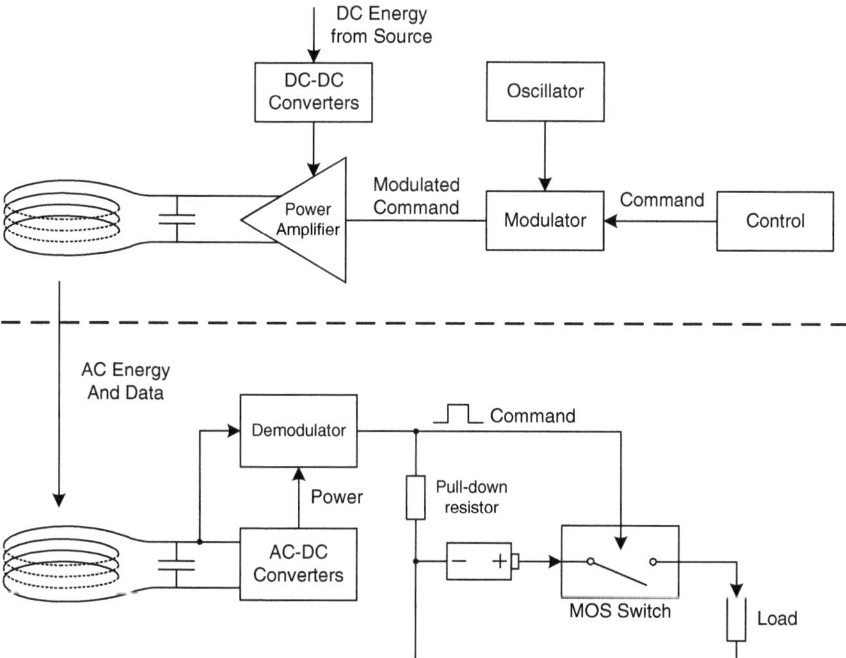

Fig. 5.16 The wireless power switch for implantable battery

modulated by the modulator. The modulated command is sent to the receiver via the power amplifier and antennas. The receiver uses the received power to demodulate and decode the command. If the command is confirmed, it triggers the MOS switch and the implantable battery is switched on. The power consumption of this system is nearly zero except the current leakage of the transistor switch, which is typically less than 10 nA.

5.6 Recharging Management

Some biomedical applications use the wireless power transfer as real-time energy, while some others use the wireless power to charge batteries. For example, the cardiac pacemaker [23, 24] requires replacing the battery for about every 5–10 years. Replacing the battery actually needs surgery, which makes patients suffering, especially because most of the patients are old people. If the wireless power transfer can be adopted to wirelessly charge the battery, maybe about every 5 years, it is a big progress for the products and a big favor for the patients. Similar, there are many other biomedical applications that need the wireless power transfer to recharge the battery. Because lithium polymer battery has relative higher energy density, we introduce the wirelessly charging and discharging circuits for the lithium polymer batteries.

As the lithium polymer battery has relative high energy density, there are some safety issues for the battery. A typical Li+ battery for cell phone stores an energy about 100 kJ, and a hand grenade has 600 kJ [25]. The energy stored in the battery is large enough to damage implantable devices and patients.

(a) Charging approach

When we are recharging a battery, we tend to make the charging time as short as possible. However, the safety of the battery must be ensured. Accordingly, the famous "constant current, constant voltage (CSCV) approach" is used in most of charging circuits. Figure 5.17 shows the CSCV approach.

As Fig. 5.17 shows, the recharging process starts with the constant current mode. The constant value of the charging current is actually the largest allowed input current of the battery. Once the battery voltage reaches 4.2 V, the charging circuit enters into the constant voltage mode. The charging circuit maintains the charging voltage at 4.2 V. Accordingly, the current diminishes with time. The charging circuit needs to monitor the temperature of the battery all the time to prevent over heat or over clod.

Discharging a battery should consider some safety issues as well. The first is the discharging circuit should be limited. The second issue is the discharging circuit should monitor the battery voltage to prevent over-discharge. Over discharging a battery would damage the chemical characteristic of the battery and the battery cannot be recharged anymore. Consequently, there is always about 20 % the

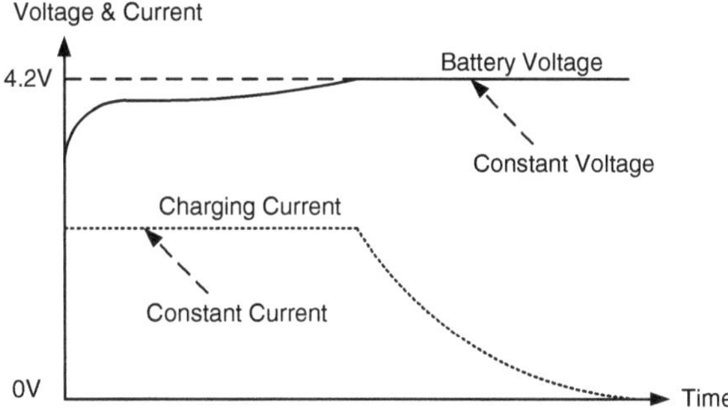

Fig. 5.17 The constant current and constant voltage approach

energy of the battery cannot be used anyway. The third issue is the discharging circuit should monitor the battery temperature.

(b) Charging circuits

A charging circuit is designed by us as Fig. 5.18 shows. The wireless power charger is composed of a power receiver, a bias generator, a temp monitor, and the main charging control circuit with many comparators and operational amplifiers.

As Fig. 5.18 shows, the power receiver converts the received AC energy to the DC power. The bias generator provides a reference voltage and bias currents for all other analog circuits in the system. The main charging circuit is made of comparators and amplifiers. The main charging circuit includes two functional blocks. One is used to monitor and regulate the charging voltage. The other is used to monitor and regulate the charging voltage. When the battery voltage is below 3.3 V, the circuit enters into the 10 % constant current charging mode, which is designed for the protection to over-discharged battery. One operational amplifier is responsible to regulate the charging current to be only 10 % of the maximal charging g current. When the battery voltage reaches 3.3 V, the current enters into the 100 % constant current charging mode, which is also called the fast charging mode. When the battery voltage reaches 4.2 V, the circuit automatically switches to the constant voltage charging mode. The corresponding amplifier ensures the charging voltage will not exceed 4.2 V.

In the wireless power transfer, the received power might not be stable, which is the most important difference to the regular charging circuit. Accordingly, in the circuit, the charging current does not fulfill the rigorous CSCV approach. The approach of this circuit is to charge all the current into the battery as long as the current does not exceed the current limitation in the CSCV approach. For example, in the constant current mode, the charging current limitation is set to be 100 mA. The charging circuit would allow any current that is less than 100 mA charging into the battery. The current limitation can be adjusted by the resistor voltage divider. The charging circuit also employs a monitoring circuit to detect the battery temperature. Once the battery is over heat or over cold, the charging process would be stopped.

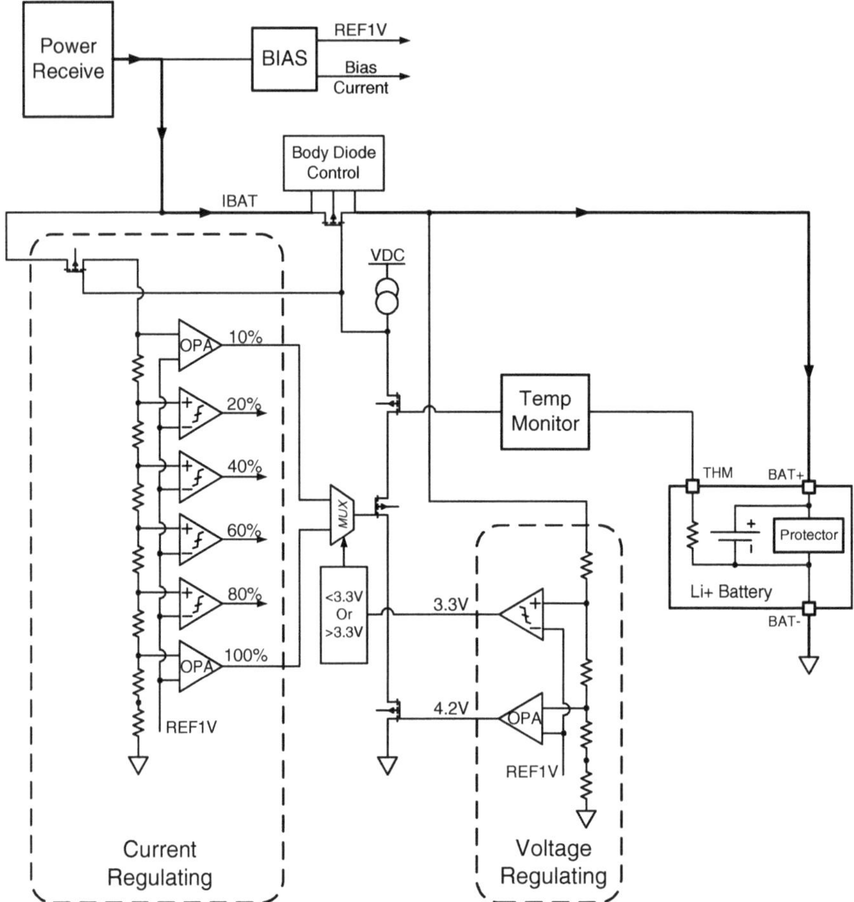

Fig. 5.18 A wireless power charger

5.7 Summing Up

In this chapter, we have presented the techniques of the power management in the wireless power transfer for biomedical microsystems. First, we have discussed the motivation of developing the power management in wireless power systems. The primary targets of the power management are to promote transfer efficiency and the system reliability. Second, a circuit model has been given at the beginning of this chapter. The model includes several kinds of power converters and the power management modules. The power management modules could have connection with all other circuits in the system. At last, several designs of the power management have been presented in detail.

According to the circuit model and the motivations, we classified the circuits of the power management into 5 designs. They are the tuning adjustments, the power regulations, the wireless power watchdog, the wireless power switch, and the recharging management circuit. The tuning adjustment is designed for automatically optimizing the resonance status. We have totally introduced 4 types of tuning adjustments. They are quite effective to eliminate the deviation of the value of the tuning components, reduce the interference caused by unexpected magnetic or metal objects between the antennas, and increase system efficiency. Another introduced power management is the power regulation. By using the power regulation, the transmitter could work only when the receiver is detected, the received power level can also be regulated to a relatively stable level, and the receiver could also negotiate power with the transmitter. The wireless power watchdog has been proposed to ensure the system reliability. It monitors the system voltage and triggers the reset of the whole system in the worst case. The wireless power switch has been presented to address a universal problem in the biomedical applications, which is how to switch on the battery in waterproofing microsystems. By using the proposed wireless power switch, the power leakage is nearly zero. The last power management we have introduced is the recharging management, which basically follows the constant-current constant-voltage approach.

References

1. Wang, G., Liu, W., Sivaprakasam, M., et al. (2005). Design and analysis of an adaptive transcutaneous power telemetry for biomedical implants. *IEEE Transactions on Circuits and Systems I: Regular Papers, 52*(10), 2109–2117.
2. Torres, E. O., & Rincón-Mora, G. A. (2009). Electrostatic energy-harvesting and battery-charging CMOS system prototype. *IEEE Transactions on Circuits and Systems I: Regular Papers, 56*(9), 1938–1948.
3. Tomita, K., Shinoda, R., Kuroda, T., et al. (2011). 1 W 3.3 V-to-16.3 V boosting wireless power transfer circuits with vector summing power controller, A-SSCC (pp. 177–180).
4. Dolgov, A., Zane, R., & Popovic, Z. (2010). Power management system for online low power RF energy harvesting optimization. *IEEE Transactions on Circuits and Systems I: Regular Papers, 57*(7), 1802–1811.
5. Maurath, D., Becker, P. F., Spreemann, D., et al. (2012). Efficient energy harvesting with electromagnetic energy transducers using active low-voltage rectification and maximum power point tracking. *IEEE Journal of Solid-State Circuits, 47*(6), 1369–1380.
6. Chen, J.-J., Lin, M.-S., Lin, H.-C., et al. (2008). Sub-1 V capacitor-free low-power-consumption LDO with digital controlled loop, APCCAS (pp. 526–529).
7. Lee, R., Ibrahim, T. S., Baertlein, B. A., et al. (2001). RF coil modeling and analysis in high field MRI: lessons learned. *Antennas and Propagation Society International Symposium IEEE, 1*, 362–365.
8. Acar, M., Annema, A.-J., & Nauta, B. (2007). Analytical design equations for class-E power amplifiers. *IEEE Transactions on Circuits and Systems I, 54*(12), 2706–2717.
9. O'Driscoll, S., Poon, A., & Meng, T. H. (2009). A mm-sized implantable power receiver with adaptive link compensation, ISSCC (pp 294–295).
10. Si, P., Hu, A. P., Malpas, S., et al. (2008). A frequency control method for regulating wireless power to implantable devices. *IEEE Transactions on Biomedical Circuits and Systems, 2*(1), 22–29.

11. Huang, W.-J., Chen, C.-L., & Liu, S.-I. (2009). A wireless power telemetry with self-calibrated resonant frequency, VLSI (pp. 80–83).
12. Kurs, A., Karalis, A., Moffatt, R., et al. (2007). Wireless power transfer via strongly coupled magnetic resonances. *Science, 317*(5834), 83–86.
13. Imura, T., Okabe, H., Uchida, T., et al. (2009). Study on open and short end helical antennas with capacitor in series of wireless power transfer using magnetic resonant couplings, IECON (pp. 3848–3853).
14. Yoo, J., Yan, L., Lee, S., et al. (2010). A 5.2 mW self-configured wearable body sensor network controller and a 12 μW 54.9 % efficiency wirelessly powered sensor for continuous health monitoring system. *IEEE Journal of Solid-State Circuits, 45*(1), 178–188.
15. Lee, S. B., Lee, H.-M., Kiani, M., et al. (2010). An inductively powered scalable 32-channel wireless neural recording system-on-a-chip for neuroscience applications. *Biomedical Circuits and Systems, IEEE Transactions on, 4*(6), 360–371.
16. Chae, C.-S., Le, H.-P., Lee, K.-C., et al. (2009). A single-inductor step-up DC–DC switching converter with bipolar outputs for active matrix OLED mobile display panels. *Solid-State Circuits, IEEE Journal of, 44*(2), 509–524.
17. Kim, N. Y., Kim, K. Y., Choi, J., et al. (2012). Adaptive frequency with power-level tracking system for efficient magnetic resonance wireless power transfer. *Electronics Letters, 48*(8), 452–454.
18. Fu, W., Zhang, B., & Qiu, D. (2009). Study on frequency-tracking wireless power transfer system by resonant coupling, IPEMC (pp. 2658–2663).
19. Cannon, B. L., Hoburg, J. F., Stancil, D. D., et al. (2009). Magnetic resonant coupling as a potential means for wireless power transfer to multiple small receivers. *IEEE Transactions on Power Electronics, 24*(7), 1819–1825.
20. Chiu, H.-W., Lin, M.-L., Lin, C.-W., et al. (2010). Pain control on demand based on pulsed radio-frequency stimulation of the dorsal root ganglion using a batteryless implantable CMOS SoC. *IEEE Transactions on Biomedical Circuits and Systems, 4*(6), 350–359.
21. Chen, J.-J., Lin, M.-S., Lin, H.-C., et al. (2008). Sub-1 V capacitor-free low-power-consumption LDO with digital controlled loop, APCCAS (pp. 526–529).
22. Kim, Y.-I., & Lee, S.-S. (2012). Fast transient capacitor-less LDO regulator using low-power output voltage detector. *Electronics Letters, 48*(3), 175–177.
23. Lee, S.-Y., Su, M. Y., Liang, M.-C., et al. (2011). A programmable implantable microstimulator SoC with wireless telemetry: Application in closed-loop endocardial stimulation for cardiac pacemaker. *IEEE Transactions on Biomedical Circuits and Systems, 5*(6), 511–522.
24. Wong, L. S., Hossain, S., Ta, A., Edvinsson, J., et al. (2004). A very low-power CMOS mixed-signal IC for implantable pacemaker applications. *IEEE Journal of Solid-State Circuits, 39*(12), 2446–2456.
25. TNT equivalent, Wikipedia, http://www.wikipedia.org.

Chapter 6
Design Cases

Abstract In this chapter, we will propose two design cases. They both are target to the wirelessly powered capsule endoscopy. The wireless power transfer helps improving the performance of the capsule endoscopy in terms of operating time and number of images. Comparing to other implantable devices like neural recorders and electrical stimulators, the capsule dynamically moves in digestive track. So, the transfer efficiency results in a broad range. In other words, the batteryless capsule endoscopy is actually a relatively complex biomedical application. The two design cases proposed in this chapter have different features and design considerations. In the first case, the wireless power is transferred from a floor to the capsule to allow patients walking freely in an inspection room. In the second case, the wireless power is transferred from a specially designed jacket to the capsule to allow patients going back home comfortably.

6.1 Introduction

In today's capsule endoscopes, the limited energy budget of batteries severely limits the system performances in terms of operating time, image resolution, and number of images [1, 2]. Furthermore, future capsule endoscopes are expected to perform even more complicated and energy-consuming tasks such as locomotion [3, 4]. Therefore, significant efforts [4–16] have been dedicated to researching an alternative energy-supply option. That is the wireless power transfer, in which power is delivered to the capsule from a remote power source. Available works on the batteryless capsule endoscope can be classified into 4 directions. They are system level designs [5–8], circuit designs [9–11], coil optimizations [12, 13], and theory analysis [14–16]. However, the wireless power based endoscopes are still facing serious challenges.

T. Sun et al., *Wireless Power Transfer for Medical Microsystems*,
DOI: 10.1007/978-1-4614-7702-0_6,

In this chapter, we clarify these challenges and present two wireless power design cases to meet these challenges. The two cases are designed with respect to different systematic level considerations. For example, one system transfers power from a floor to the capsule to allow patients walking freely in an examination room, and the other transfers power from a specially designed jacket to the capsule to allow patients going back home comfortably.

In each of the case, we firstly introduce the motivation of the system. Then, the system considerations, system modeling, and system diagram are to be introduced. The antenna and circuit techniques in the previous chapters are to be adopted to meet the system's requirements. Experiments and conclusions will be given at the end. By reading this chapter and referring these designs, readers may design more advanced wireless power transfer systems for biomedical applications.

6.2 Design Case A

6.2.1 Introduction

In regular batteryless capsule endoscopy systems, patients are required to wear on a jacket with a power transmitter and transmitting antennas, while an external power source is connected to the power transmitter via a long power cable [5, 10]. As a result, the patient's activity is restricted by the power cable. It makes patients both psychologically and physically uncomfortable. Moreover, the moving power cable and its connectors have reliability problems.

To address the problem, this section presents a power transfer system for the application of the capsule endoscopy. The operation of the proposed system is illustrated in Fig. 6.1. In the proposed system, a wireless power transmitter array is installed under the floor of a testing room. As a result, wireless power can be ubiquitous in an examination room. The patient wearing a special jacket is able to walk freely in the room. The jacket is equipped with a power repeater, which is actually a resonant power antenna. Pressure sensors would identify the patient's position, so the system activates the nearest transmitter to generate wireless power. The power is delivered to the power repeater in the jacket, which would produce a relatively strong magnetic field nearby the abdomen area. The capsule inside the patient employs an efficiency-enhanced power receiver to pick up energy from the power repeater in the jacket. Comparing to existing designs, the proposed system does not restrict the patient's movement and eliminates the unreliability problem caused by moving power cables and their connectors.

The power transfer efficiency is clearly an essential problem in the system. In existing designs, the transmission distance is approximately dozens of centimeters, while the transfer efficiency is in the range of 1–33.1 % [5, 10, 14, 16]. In our system, the transfer distance from the floor to the capsule is around 1 m, which is approximately 5–30 times longer than previous results. Since the power efficiency

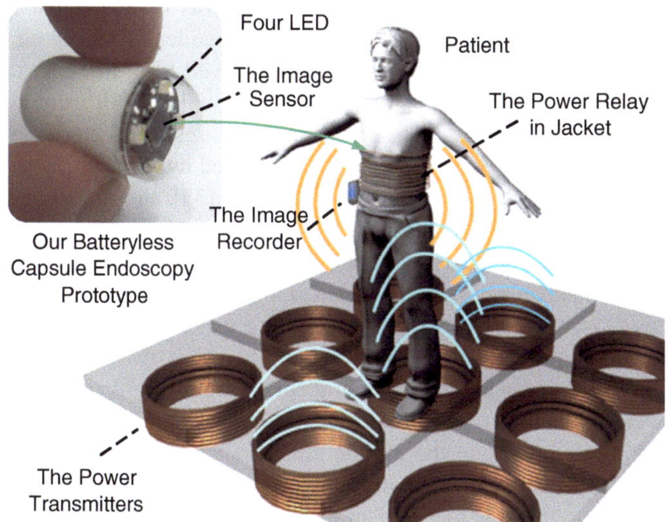

Fig. 6.1 Wireless power transmission from the floor to the capsule

follows an inverse cubed law with regard to the distance in non-radiative transfer, it is extremely challenging to maintain sufficient power efficiency in this system. The major contributions of this work can be summarized as follows.

1. We proposed a two-hop wireless power transfer link for capsule endoscopes. Taking advantage of the strong-coupling, the first hop transfers power from the power transmitter array to the jacket with high efficiency over long distance. Taking advantage of the loose-coupling, the second hop delivers the power from the jacket to the small-size capsule.
2. We developed energy-efficient circuits for enhancing power efficiency. First, a switch-mode rectifier with higher rectification efficiency is proposed. Second, a power combination circuit with higher combining efficiency is developed. These circuits can be applied to other systems such as the neural recorders [17–19], the body sensor networks [20], the implantable pain controllers [21], the sub-cubic millimeter implants [22], the implants for drug tests [23], and even the RFID applications [24].

6.2.2 The Two-Hop Transfer Link

In this section, we firstly give a brief summary to the existing wireless power transfer mechanism related to the capsule endoscopy and then introduce the details of the proposed two-hop transfer link.

Inductors and capacitors in series or parallel connections have been commonly employed as resonators for inductive power couplings. In 2007, the strongly-coupled magnetic resonance mechanism was demonstrated [25]. Since then, the wireless power transfer has attracted considerable research efforts. In 2008, the transcutaneous power transfer was improved [26]. In 2009, the loosely-coupled transfer [27–29]. The magnetic resonant coupling [30] were analyzed. In 2010, the helix transmitter [31], the evanescent resonant coupling [32], the loosely-coupled coil design [33], and the small folded cylindrical helix dipole [34] were studied. In 2011, the wireless repeater effect [35], the misalignment model [36], the maximization of air gap [37], and the design optimization for implants [38] were researched. As a brief summary, according to the coupling strength, the existing WPT mechanisms can be divided in two groups, the loose-coupling mechanisms [27–29, 33] and the strong-coupling mechanisms [25, 30, 31, 34, 35, 38]. The loose-coupling mechanisms commonly adopt two coils as resonators. They are the transmitting and the receiving coils. The strong-coupling mechanisms commonly employ a four-coil topology [25, 30] or a many-coil topology [35]. The strong-coupling is the key to the middle range transfer.

According to the specific demand of the target application, in this work we propose a two-hop transfer mechanism as shown in the Fig. 6.2. In the first hop, the strong-coupling with high-Q resonators are adopted. An end-fire helix is employed as the power transmitter under floor, while an open-ends helix is employed as the power repeater in the patient's jacket. In the second hop, the loose-coupling with small size receiving antenna is designed. Small size regular LC resonators are deployed as the receiving antennas in the capsule.

The proposed mechanism is a result of a full consideration of the power efficiency and the size limitation on the capsule. To deliver power from the floor to the power repeater, a middle range transfer is required. Accordingly, we deploy the

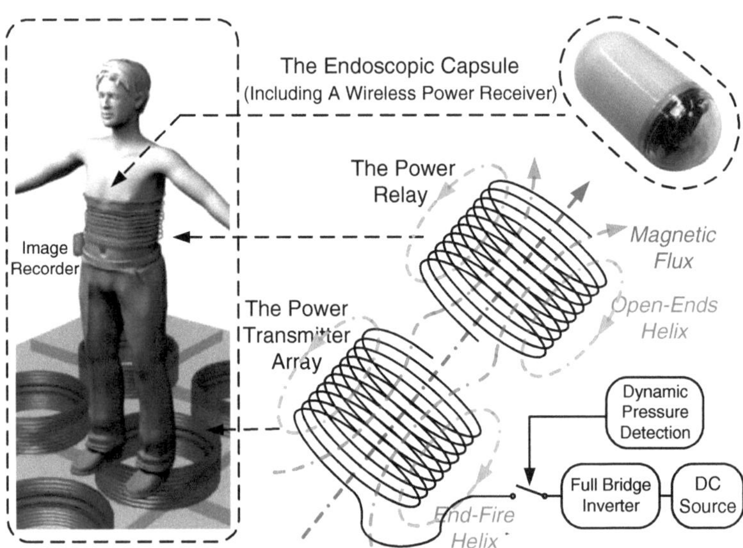

Fig. 6.2 The two-hop wireless power transfer mechanism

strong-coupling in the first transfer hop. Considering the capsule has very strict limitations on the size and comparing to the helixes, the LC resonator has much smaller size at the same operating frequency, the loose-coupling with LC resonators are used in the capsule. This mechanism combines the strong and the loose couplings together to transfer the power to the small size receiver over a middle range transmission distance.

To precisely describe the transfer performance of this structure, we cite the Eqs. 3.34 to 3.40. For convenience, these Equations are given as follows. First, the current I_{Load} in the capsule is:

$$I_{Load} = \frac{U_s \cdot \omega^2 M_{12} M_{23}}{(R_S + R_1)R_2(R_3 + R_L) + \omega^2 M_{12}^2 (R_3 + R_L) + \omega^2 M_{23}^2 (R_S + R_1)} \quad (6.1)$$

where R_S, R_1, R_2, and R_3 are the equivalent internal resistances of the AC power source, the transmitter under floor, the power repeater in the jacket, and the LC resonator in the capsule respectively. The variable R_L represents the resistance of the load. The variables M_{12} and M_{23} are the mutual inductances between the transmitter, the power repeater, and the capsule. The variable U_s is the voltage amplitude of the AC power source, and the ω is the resonant angular frequency. The following Eq. 6.2 shows the definitions of the coupling factors and the Eq. 6.3 shows the definitions of quality factors.

$$M_{12} = k\sqrt{L_1 L_2}, \quad M_{23} = k\sqrt{L_2 L_3} \quad (6.2)$$

$$R_1 = \frac{\omega L_1}{Q_1}, \quad R_2 = \frac{\omega L_2}{Q_2}, \quad R_3 = \frac{\omega L_3}{Q_3} \quad (6.3)$$

where Q_1, Q_2, and Q_3 are the Q factors of the resonators, and the K_{12} and K_{23} are the coupling factors between them. If the coupling factors and quality factors are used to represent the current on the load in capsule, the Eq. 6.1 can be donated as Eq. 6.4.

$$I_{Load} = \frac{U_s \cdot \omega^2 k_{12} k_{23} L_2 \sqrt{L_1 L_3} \cdot Q_1 Q_2 Q_3}{\omega^3 L_1 L_2 L_3 + \omega^3 k_{12}^2 Q_1 Q_2 L_1 L_2 L_3 + \omega^3 k_{23}^2 Q_2 Q_3 L_1 L_2 L_3} \quad (6.4)$$

If the impedance matching condition $k_{12}^2 Q_1 = k_{23}^2 Q_3$ is satisfied, and if we define factor X, the Eq. 6.4 has its optimal value as Eq. 6.5 expressed.

$$I_{Optimal} = \frac{X}{1 + 2X} \cdot \frac{U_s}{\sqrt{(R_S + R_1)(R_3 + R_L)}}, \quad X = k_{12} k_{23} \cdot Q_2 \sqrt{Q_1 Q_3} \quad (6.5)$$

To maximize the K_{23}, the diameter of the receiving coils in the capsule should be optimized as large as possible. In our final design, the diameter is 11 mm because the diameter of the capsule is 13 mm. The dimension of the power repeater is 21 by 31 cm, which is decided by the patient's body size. The diameter of the transmitter is 48 cm, which is determined by maximizing K_{12} [39, 40]. Besides the coupling factors, the Q factors are also critical. As shown in the

Eq. 6.5, the Q_2 has a large impact on the efficiency. High Q open-ends helical antennas are employed as the power repeater and the transmitter to improve the Q_2 and Q_1. They significantly improve the power transfer efficiency, especially for the first hop transfer. The quality factor Q_3 of the receiving coil has the lowest value because the capsule's size is too limited.

What operating frequency should this transfer structure works at? In a previous research on the strongly coupled magnetic resonances [25], the frequency of 10.56 MHz was given as a frequency for antennas with the radius of 30 cm and the transfer distance in range from 0 to 2.25 m [25]. Considering this work has similar geometrical parameters, we adopted and verified the nearest ISM frequency, 13.56 MHz. Experiments confirmed that the frequency of 13.56 MHz is a qualified choice for our system.

Figure 6.3 shows the system under the floor. They are the pressure sensors, transmitters, and helical antennas. Pressure sensors determine the working status of the transmitters, while the transmitters are actually H-bridge inverters. Their outputs are connected to the helical antenna using high frequency transformers.

To ensure the wireless power transfer is continuous, transmitting strategies are proposed. Pressure sensors under the floor find out the patient's position. At most time, only one transmitter embedded in the floor will be activated.

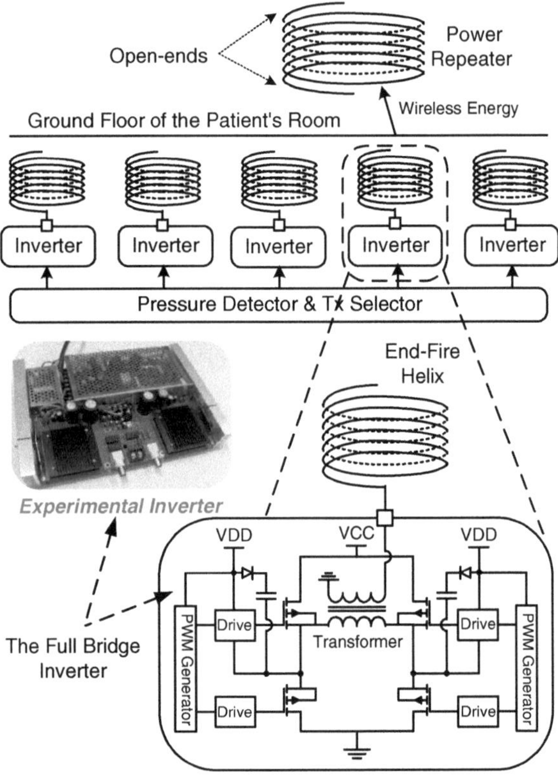

Fig. 6.3 Transmitters *under* the floor

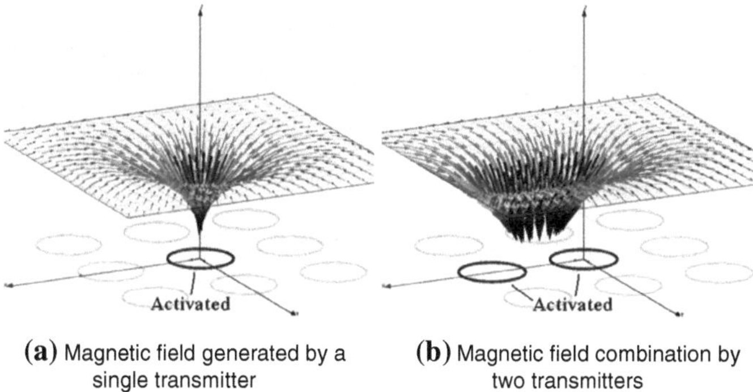

(**a**) Magnetic field generated by a single transmitter

(**b**) Magnetic field combination by two transmitters

Fig. 6.4 Magnetic field on a virtual plane 1 m *above* the floor to indicate the magnetic field strength on the patient

However, when a patient is walking between two transmitters, the switching operation of the transmitters is required. To ensure the power transfer is continuous, the first transmitter is turned off after the second transmitter is turned on. The two transmitters work simultaneously for a short period of time. As Fig. 6.4 shows, it is assumed there are totally nine transmitters in an array under the floor. They have the identical diameter of 48 cm and are regularly placed with a center-to-center distance of 100 cm. In Fig. 6.4, virtual planes 1 m above the floor are used to indicate the magnetic field on the patient. When the patient is above one transmitter, the magnetic vector generated by a single transmitter is shown in Fig. 6.4a. The magnetic field reaches its maximal value in the axial direction. When the patient is between two transmitters, the magnetic vector generated by two transmitters is shown in Fig. 6.4b. It is clearly that if the two currents in the transmitters flow in the same direction (realized by sharing global clock among transmitters), the magnetic fields above 1 m are superposed.

How to ensure the stability of the transfer? When a patient walks around above two transmitters, our system uses two techniques to ensure the stability of the system. (1) By using the strong-coupling technique in the first transfer hop, the power transfer efficiency does not change significantly with regard to the patient's position. The system remains relative high transfer efficiency in the worst case. The detailed measured coupling efficiency is given in the experiments. (2) When the patient walks in the middle of two transmitters, the system may repeatedly switch transmitters. In order to prevent switching too often, all the switching operations of the transmitters in the system are Schmitt triggered.

6.2.3 The Efficiency-Enhanced Receiver

The overall power efficiency is determined by the antenna coupling efficiency and the circuit conversion efficiency. Therefore, the design of the power receiver in the

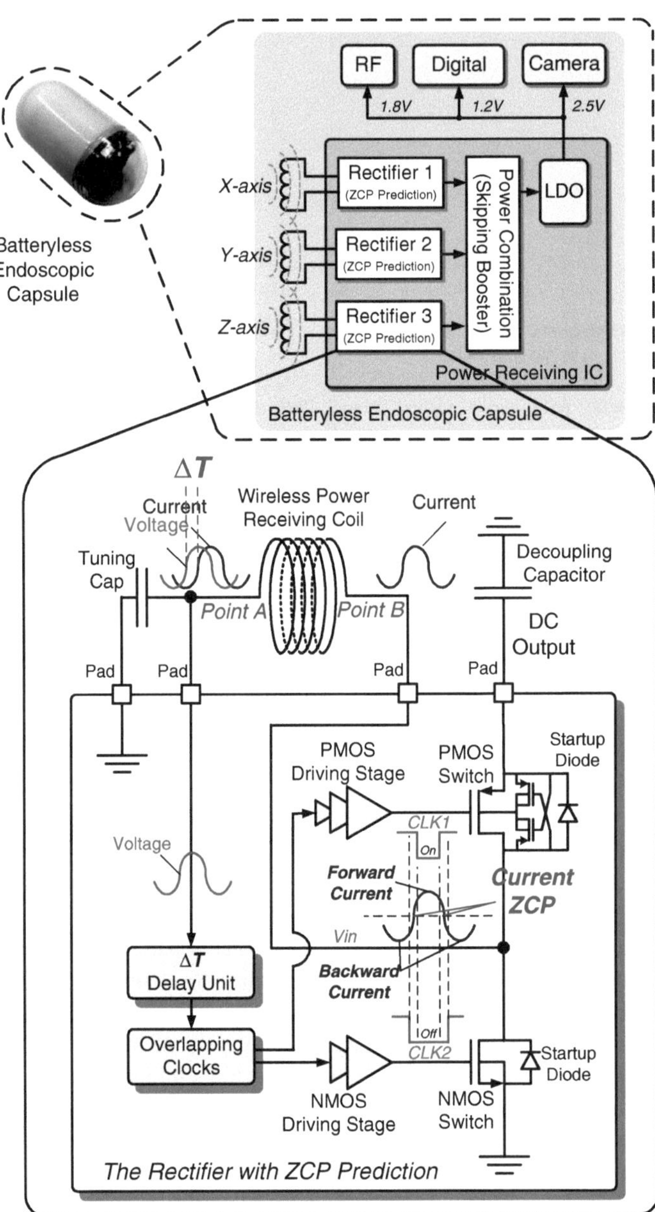

Fig. 6.5 The circuit diagram of the endoscopic capsule

capsule is also critical. In this section, we presented the detail design of the circuit in the capsule. Figure 6.5 shows the circuit diagram of the power receiver. In the figure, the adopted rectifiers with ZCP prediction proposed in Chap. 4 are adopted and emphasized.

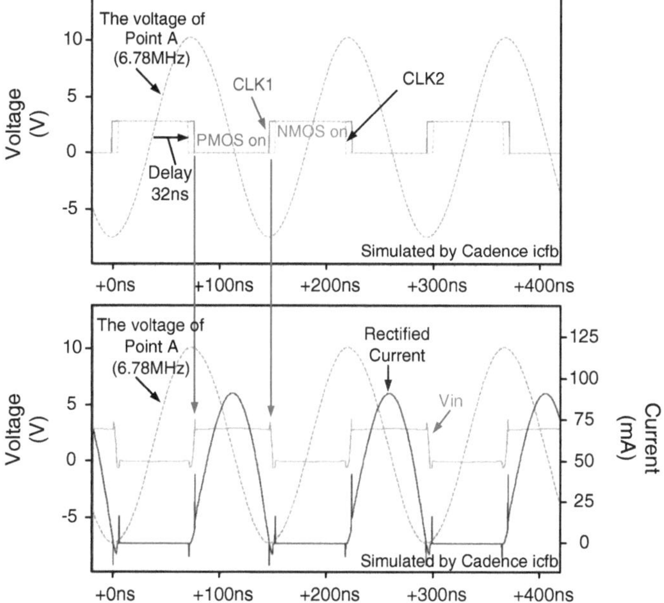

Fig. 6.6 Simulated waveforms in the rectifier with ZCP prediction

As Fig. 6.5 shows, the efficiency-enhanced power receiver in this work adopts three rectifiers with ZCP prediction and a power combination circuit. We firstly introduce the rectifier, then the power combination circuit.

1. The rectifier with ZCP prediction

As introduced in Chap. 4, the essential of the switching timing of the rectification is to switch-on all forward current and switch-off all backward current. Once the switching timing is precise enough, the rectification efficiency can be enhanced. According to this idea, we proposed a CMOS switch-mode rectifier with current zero-crossing-point (ZCP) prediction [9].

The essential idea of this rectifier is to predict the coming of the AC current by delaying the voltage in the LC resonator. It is well-known that a fixed phase difference ($\pi/2$) exists between the voltage and the current signals in the LC resonant circuit. Therefore, we could detect the voltage ZCP ahead to predict the coming of the current ZCP behind. Accordingly, the switching timing can be precisely controlled. Figure 6.5 shows the circuit and Fig. 6.6 shows the waveforms.

These waveforms include input AC signals, the *CLK1*, the *CLK2*, the voltage of *Point A*, and the rectified current. The voltage of *Point A* is a sinusoidal wave. The rectifier delays the wave approximately 32 ns (for 6.78 MHz AC input as an example) and generates two overlapping digital waves *CLK1* and *CLK2*. They drive the PMOS and the NMOS transistors. As a result, the current is precisely rectified. As Fig. 6.6 shows, the rectified current through PMOS is the upper half of the sine wave.

2. The power combination circuit

An omnidirectional power receiver is necessary for the system because the orientation of the capsule is random. In existing designs [4–6], two or three dimensional coils are connected to multiple rectifiers. The output voltages of the rectifiers are determined by the capsule's angle to the magnetic line. The outputs of the rectifiers are connected together to combine the power together. However, when the induced voltages in the coils are significantly different, the voltage at the common output point will be determined by the rectifier with the highest DC output voltage. The other rectifiers with relative low output voltage will be back biased. Although energy is received in other two rectifiers, they cannot be delivered to the load. The power combining efficiency consequently decreases. This phenomenon is very similar to new and old batteries are mixed together.

To address this problem, this work presents the Skipping Booster as Fig. 6.7 shows. The power receiving antenna is commonly LC resonant circuits. As shown in Fig. 6.7, if the received energy is not converted from the LC circuit to the load

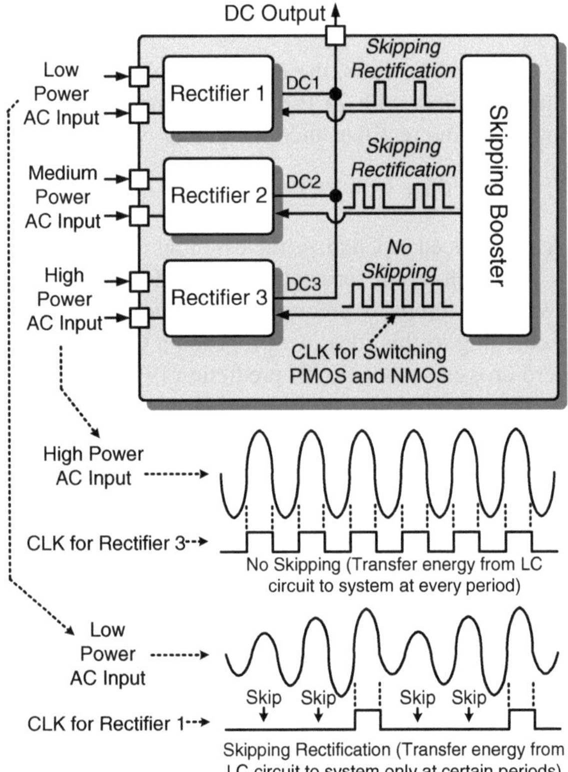

Fig. 6.7 The working principles of the power combination circuit or the skipping booster

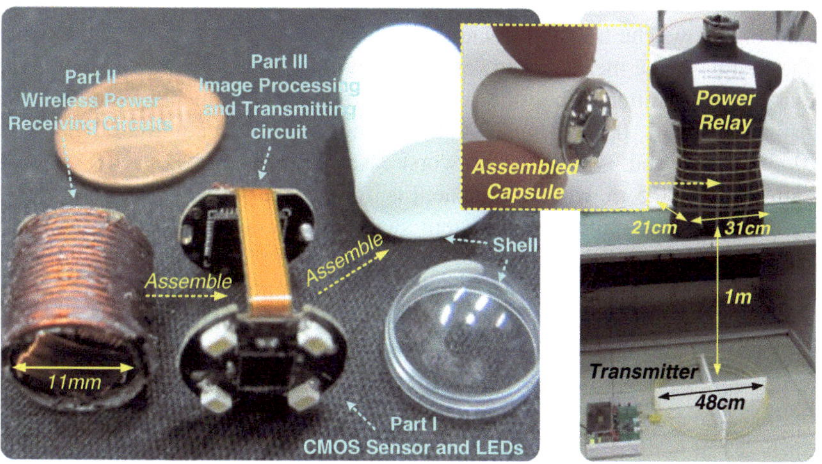

Fig. 6.8 The experiment of the wirelessly powered capsule prototype

in every cycle, the AC voltage in the LC circuit will grow cycle by cycle. In other words, the Skipping Booster boosts the AC input voltage of the rectifiers by skipping some of the rectification cycles. Once the voltage is high enough, the rectifier delivers its energy to the load. As a result, more energy can be utilized and the efficiency is promoted.

In the Fig. 6.7, suppose the rectifier 1 has the lowest input power when the rectifier 3 has the highest input power. Accordingly, some rectification cycles of the rectifier 1 are skipped and the voltage in the LC circuit 1 increases cycle by cycle. Such a working principle converts the low AC input voltage of the rectifier 1 to a high level. As a result, the rectifier 1 outputs DC power but only in some cycles. The detection of the input voltage is fulfilled by comparators and the whole circuit needs no off-chip component. The efficiency improvement of the Skipping Booster varies at different capsule's angle to the magnetic line.

6.2.4 Experiments

A prototype of the wirelessly powered capsule endoscope is implemented. As shown in Fig. 6.8, a human body model on table filled with saline water is installed to simulate the human body environment. The power repeater is winded on the human model. On the left, the capsule employs three parts of circuits. The first part is a CMOS sensor and four LEDs for imaging. The second part is the power receiver including three dimensional coils. The last part is an image processing circuit.

First, using the system prototype, the strongly coupling in the first hop is tested. Figure 6.9 shows the transfer efficiency with regard to the position of the patient. In the best case, the measured power efficiency is 39.8 % when the patient is right above one transmitter. When the patient's lateral misalignments to the activated transmitter are 0.1, 0.2, 0.3 and 0.4 m, the measured AC coupling efficiency are 38.0, 36.7, 35.5 and 34.7 %. In the worst case, the efficiency is 33.1 % when the misalignment is 0.5 m, where the patient is in the center of the two transmitters. If the patient goes on moving, the system switches to another transmitter.

Second, the overall two-hop AC–AC coupling efficiency is measured. In the base case, the coupling efficiency from the transmitter under the floor to the capsule is 5.2 % when the transfer distance is 1 m. The conditions of the best case are the lateral misalignment between the transmitter and the power repeater is zero and the capsule is positioned near the power repeater. In the worst case, the coupling efficiency is 0.3 %. The worst case appears when the lateral misalignment is 0.3 m and the capsule is positioned in the center of the human body. The domain reason for the efficiency decline is the second hop.

Third, the efficiency-enhanced power receiver is measured. Figure 6.10 shows the experimental results. Both the input and output power vary with the load resistance. The measured highest rectifying efficiency is 93.6 % during the output is 88 mW. Comparing to previous rectifiers [17, 20] for implants, this work enhances the rectifying efficiency by 13.4 %.

Forth, the efficiency enhancement of the Skipping Booster is measured. The power receiver with three dimensional coils is placed in an AC magic field and rotated over angles to simulate the random rotation of the capsule in human body.

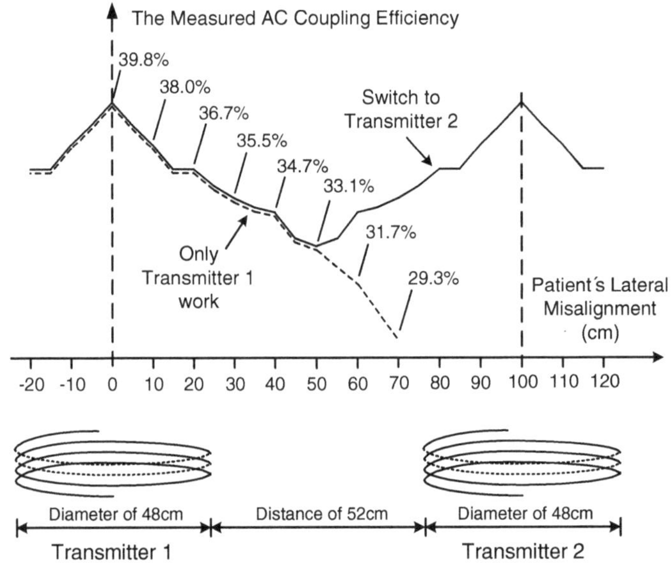

Fig. 6.9 Measured AC power coupling efficiency with regard to the position of a patient

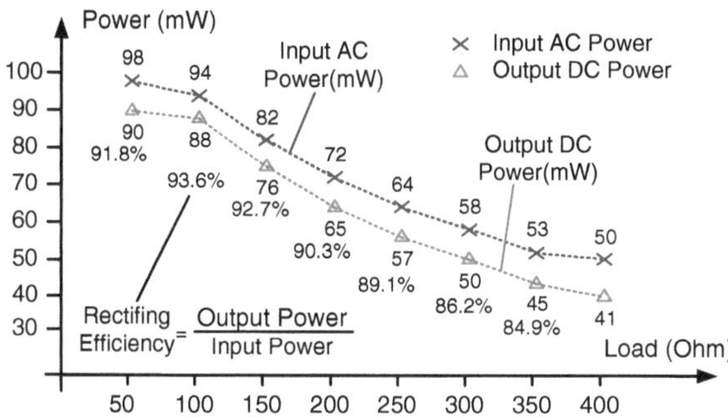

Fig. 6.10 The measured efficiency of the rectifier with ZCP prediction

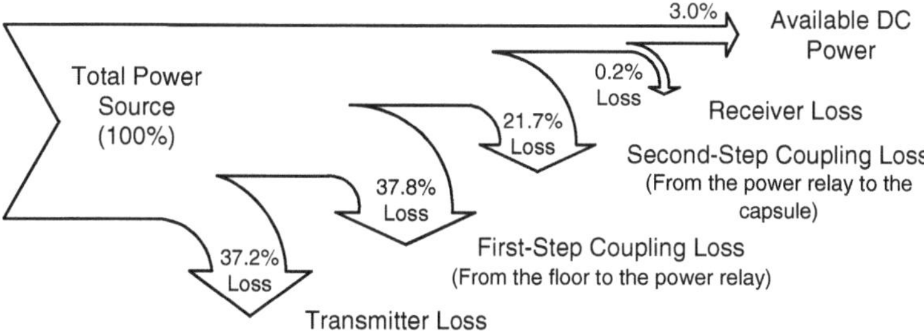

Fig. 6.11 The overall DC-to-DC transfer analysis of the system

Due to the capsule is an axisymmetric shape, only the angle contained by the magnetic line and the axis of the capsule is rotated (from 0° to 180°). The Skipping Booster is activated and deactivated to measure the efficiency enhancement. In the best cases, the efficiency improvement is 18 % at the angles from 25° to 35° or from 145° to 155°. In the worst cases, there is no improvement at the angles around 0°, 45°, 90°, 135°, and 180°. In these angles, the AC voltages on the coils are either the same or zero, thus it does not improve the efficiency.

The overall DC–DC power efficiency of the system is measured and illustrated in Fig. 6.11. In the best case, 37.2 % of the power is wasted by the transmitter, 37.8 % is lost at the first transfer hop from the transmitter to the power repeater, 21.7 % is dissipated at the second transfer hop from the power repeater to the capsule, 0.2 % is missed by the power receiver in the capsule, and finally 3.0 % of the total power can be converted to available DC power for the capsule. According to Table 6.1, comparing to existing systems [5, 10, 14, 16], the maximum overall DC–DC power efficiency of 3.0 % is a remarkable result because the transfer distance in this system is 5–30 times enlarged and the power efficiency should be in inverse cubed law of the distance.

Table 6.1 Comparison with other batteryless capsule systems

Systems	Tx/Rx diameter (cm)	Coupling type	Freq. (Hz)	Power efficiency
Reference [5]	41/0.1	Loosely	1.0 M	DC efficiency of 1 % @ 20 cm
Reference [10]	NA/NA	Loosely	1.0 M	DC efficiency of 1.8 % @ 20 cm
Reference [14]	30/0.6	Strongly	8.2 M	AC efficiency of 26.1 % @ 3 cm
Reference [16]	20/0.5	Loosely	0.5 M	AC efficiency of 33.1 % @ 10 cm
This work	48/0.11	Mixed	13.56 M	DC efficiency of 3.0 % @ 100 cm

6.3 Design Case B

6.3.1 Introduction

In the previous section, we have presented a wireless power transfer system to deliver power from a floor to the capsule. Although the introduced system allows patients to move freely in an examination room, the patient cannot leave the room. To make the system more convenient and comfortable, in this section, we present another design case of the wireless power transfer that could allow patients going home with a wireless power jacket. The jacket uses a group of batteries as its power source. Power circuits on the jacket convert the energy from the battery to wireless power and transfer the energy to the capsule in the patient. Because the power source is portable battery, it has stricter requirement on the power transfer efficiency. In this section, we will propose both antenna and circuit techniques for the improvement of the efficiency.

6.3.2 The Segmented Transfer Link

In this section, we propose a segmented transfer system for the endoscopic capsule. As shown in Fig. 6.12, the power system mainly consists of a group of vertically segmented transmitting coils, a control unit, a transmitter (TX), and a portable battery. All these components are integrated on a jacket.

How does this system work? According to the position of the capsule, the control unit selects the nearest transmitting antenna to work. The algorithm of the selection is based on the feedback of the power information from the capsule to the jacket by using wireless data connectivity. Because energy is delivered according to the position of the capsule, the new proposed system increases the average power transfer efficiency. By increasing the efficiency, the harmful electromagnetic radiation can be reduced. Moreover, a portable battery will be enough to power the whole system. Without the cable connected to the city power, the system becomes safer and the patient can be much more comfortable. Figure 6.13 shows the system diagram.

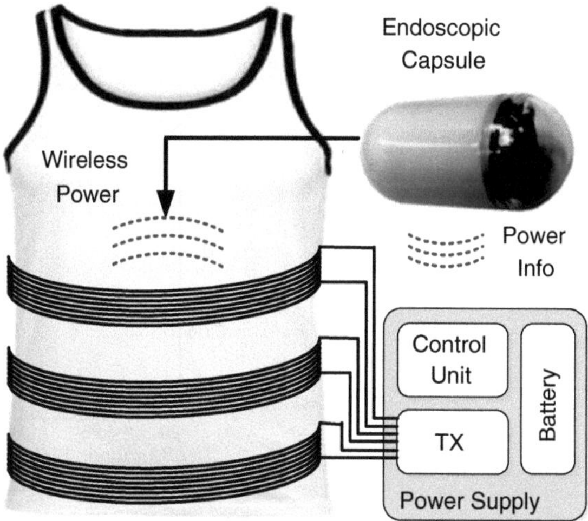

Fig. 6.12 A segmented power transmitting system for endoscopic capsule

The wireless power transmitter in the jacket consists of an inverter, a timing unit to control the inverter, a group of batteries, a multiplexer to select the segmented antennas, a RF module to receive data from the capsule, and a controller to decode data and control the multiplexer. The capsule in vivo is actually named the Micro-Ball [13]. It consists of six CMOS image sensors, six power receiving antennas, three rectifiers and a control circuit. The proposed system operates as follows. The transmitter in the jacket would select one transmitting antenna to work. If the endoscopic capsule is near the antenna selected, it picks up energy, starts to work, and sends out feedback information by using the RF module. The controller in the transmitter would listen to the feedback from the capsule. If the feedback indicates the power efficiency is high enough, the transmitter would keep

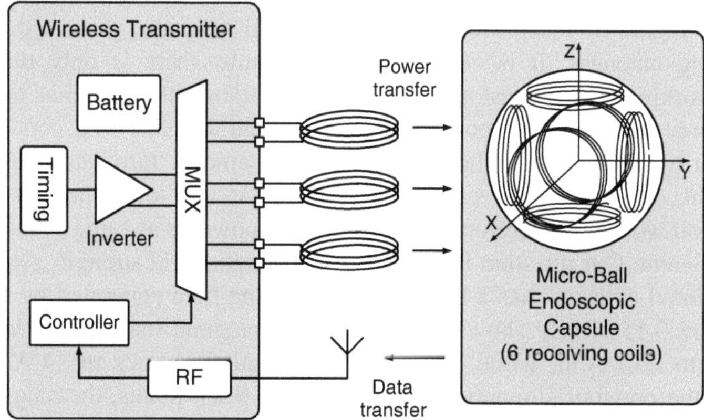

Fig. 6.13 System diagram

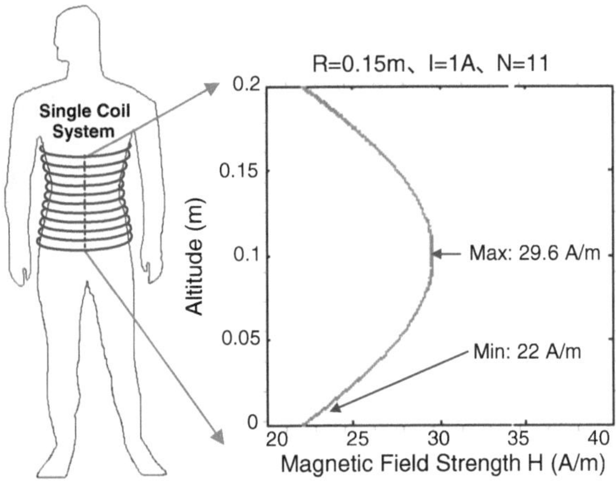

Fig. 6.14 The regular single-coil transfer system

the selected antenna working. If else, the transmitter would consider selecting another antenna to work.

Why segmented transmitting antennas increase the power transfer efficiency? We firstly explain the magnetic field generated by the regular single transmitting coil. Suppose the conventional primary coil has 11 turns and radius of 15 cm. If the current in the coil is 1 A, the magnetic field strength distribution along the axis of the primary coil can be achieved through accumulating the magnetic field generated by each turn of coil. The magnetic field strength along the axial direction of human body can be calculated by Eq. 3.20. The calculated result is showed in Fig. 6.14.

As Fig. 6.14 shows, the magnetic field strength is not uniform. The magnetic field strength has the highest value 29.6 A/m in the middle abdomen. The lowest values 22 A/m appear on the top and the button of the coil. It is clear that the capsule would have higher efficiency at the area of the middle abdomen.

In this section, our idea is to power the capsule according to its position. As a result, the transmitting system with several segmented transmitting antennas is designed. Figure 6.15 shows the magnetic field strength generated by three transmitting antennas. It is noted that at any time, there is only transmitting antenna working. The nearest antenna is the only transmitting antenna that works.

The three antennas are positioned on different heights. We conducted the simulation by assuming the three coils have the same 11 turns and the radius of 15 cm. The current in each of the coil is 1 A. At any time, there is only one selected coil working. Therefore, the transmitting power is as same as the previous last experiment. Our question is whether the magnetic field strength generated by the segmented antennas has a higher value than the field generated by the single coil. Figure 6.15 clearly shows the result. The maximal magnetic field strength increases to 36.7 A/m, while the minimal strength increases to 34 A/m. Comparing to the previous simulation, at the best and worst points, the magnetic field strength are improved by 23.9 and 54.5 % respectively.

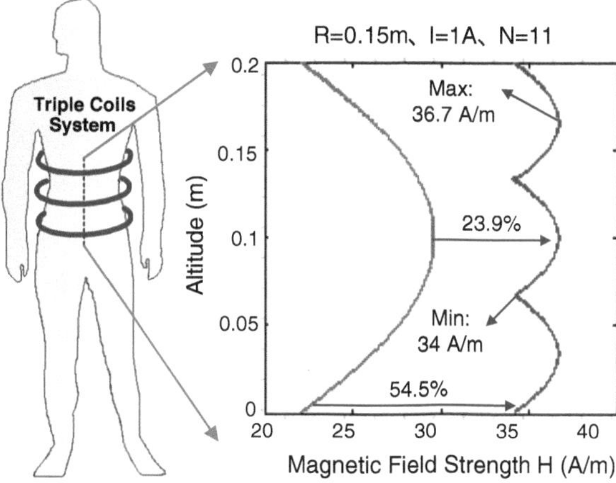

Fig. 6.15 The proposed segmented transfer system

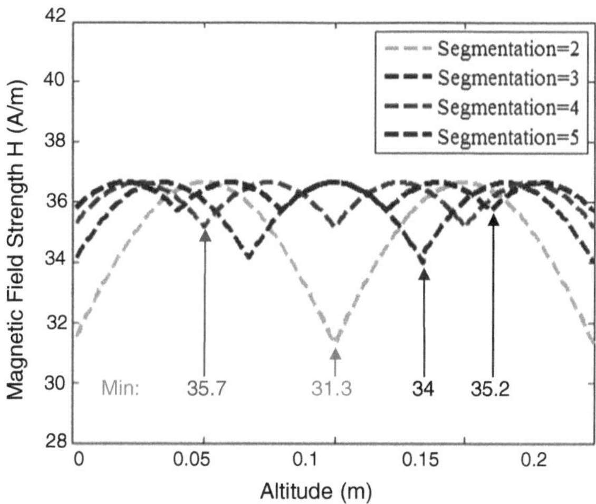

Fig. 6.16 The magnetic field strength with respect the number of segments

In the above simulation, we assume there are total three antennas. But how many antennas is the best solution? To answer this question, we conducted another simulation.

As shown in Fig. 6.16, there are four different numbers of segmentation. It's clear that the more transmitting antennas, the magnetic field becomes more uniform. When there are two antennas, the minimal magnetic field strength is 31.3 A/m. When there are three antennas, the strength becomes 34 A/m. And when there are four and five antennas, the strength is 35.2 and 35.7 A/m respectively. According to these data, we determined to use three transmitting antennas to make a balance between the transfer efficiency and the design complexity of the project.

A prototype was fabricated in Fig. 6.17. A human body model is adopted.

As the above figure shows, there are 3 transmitting antennas on the model. They are all connected to the power transmitter. The transmitter is composed of two

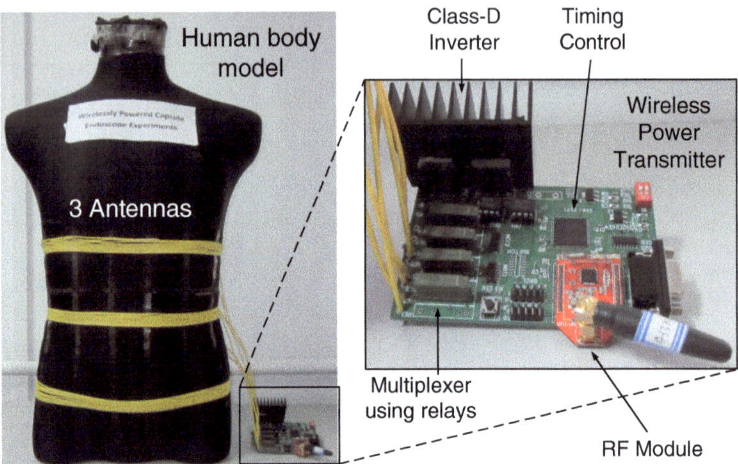

Fig. 6.17 The experiments on the segmented transmitter

power MOSFETs, a timing control part to control the inverter, a multiplexer using relays, and a RF module. The detailed experiments with receiving side will be introduced later.

To conclude this section, we have proposed a new antenna solution for the capsule endoscopy. The new solution increases the transmission efficiency by selecting the nearest transmitting antenna to transfer energy to the capsule. The maximum and minimum magnetic field strength can be improved by 23.9 and 54.5 % respectively.

6.3.3 The Full-Directional Receiver

In the Sect. 4.2.4, we have explained that the wireless power transfers usually adopt 3-dimensional receiving coils for the full directionality. The capsule endoscopy is such a typical system. In this section, we will adopt the proposed power combination circuit for the capsule. To make full use of the wireless power received by all coils, it is critical to maintain an identical output voltage over all coils. Based on such an idea, a novel omnidirectional wireless power receiving circuit is proposed and experimented. Figure 6.18 shows the circuits.

The circuits consist of six receiving coils, a control unit, a group of LDO, and three receiving channels, X, Y and Z. Each receiving coil is series connected to the coil in opposite position, so there are actually three coils, which would receive wireless energy from three different directions. Each receiving channel includes a CMOS self-synchronous full-bridge rectifier and a PWM-control boost DC–DC converter. With an adjustable voltage gain control, the boost DC–DC converter could convert a relatively low output voltage of the rectifier to a high level, so it is

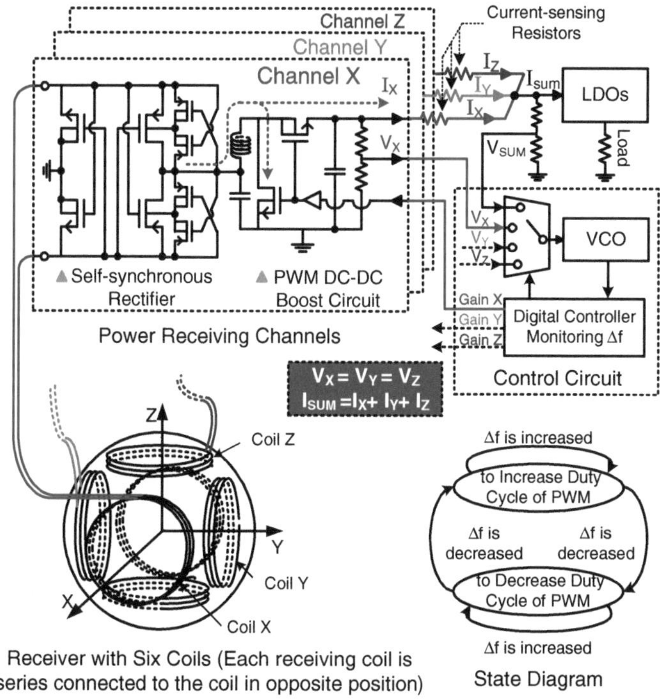

Fig. 6.18 The proposed integrated omnidirectional power receiving circuit

possible to guarantee for the three channels to deliver similar output. To keep an identical voltage output over all channels, the voltage gain of boost DC–DC converter is dynamically adjusted by the control unit. The control unit consists of three resistors, a VCO, a multiplexer, and a digital controller. The resistors are deployed at the output port of each channel. The VCO would quantize each of dropout voltages on the resistors to a value of frequency. The digital controller would measure the frequencies, calculate the output voltage of each channel, and adjust the gain of each boost DC–DC converter by changing the duty cycle of a corresponding PWM wave. Accordingly, the novel circuit could make the same output voltage over all receiving channels and effectively combine energy from all space directions. The combined energy is finally delivered to the load through LDO.

6.3.4 Experiments

The proposed receiving circuit was implemented in a 0.18 μm CMOS process, and it is bonded to six 7 × 7 × 3 mm receiving coils with three 330 pF tuning capacitors to constitute a wireless power receiver. The chip micrograph and a photo of the receiver are presented in Fig. 6.19.

Fig. 6.19 Experimental prototype of the capsule endoscopy

This receiver has a diameter of 13 mm. It is supposed to be installed into the endoscopic capsule. In order to illustrate how the proposed circuit increases the received power, the position of the receiver is firstly fixed at the central point of abdomen and the angle between the receiver's axes Z and the magnetic field is fixed at 0°, so there is little induced energy from channel Z. The angle between the receiver's axes X and the magnetic field is denoted as angle α. When the angle α varies, the induced voltages of channel X and Y change. The experiment was conducted to measure the receiving power as a function of the angle α. Only the angle α from 0° to 45° is measured because the structure of the receiving coils in our design is centre symmetric. Figure 6.20 shows the measured result.

As Fig. 6.20 shows, comparing to the previous result [5], an efficiency improvement of 16.2 % is observed at the angle α of 30°. At this angle, both channels X and Y output a reasonable large voltage but with a remarkable difference in between. In the previous work [5], the minor voltage of two channels cannot be exploited. Our circuit could effectively combine the energy from both channels. The output current of channel Y in the proposed power receiver is improved from 2.5 mA in the conventional way to 4.5 mA. The worst-case performance improvement appears at 0° and 45°. In fact, the received energy from the channel Y is zero at 0°, and the output voltages of the channels X and Y are already equal at 45°. The measured total power consumption of the resistor network, VCO and the digital controller is no more than 1 mW, which could be neglect.

The overall transfer efficiency is also measured. The adopted transmitter is the segmented transmitter with total 3 transmitting antennas introduced in previous section. The experimented transmitter is illustrated in Fig. 6.17. To simulate the human body environment, the human body model is filled with physiological saline. In the best case, a maximal power efficiency of 5.4 % is observed when the

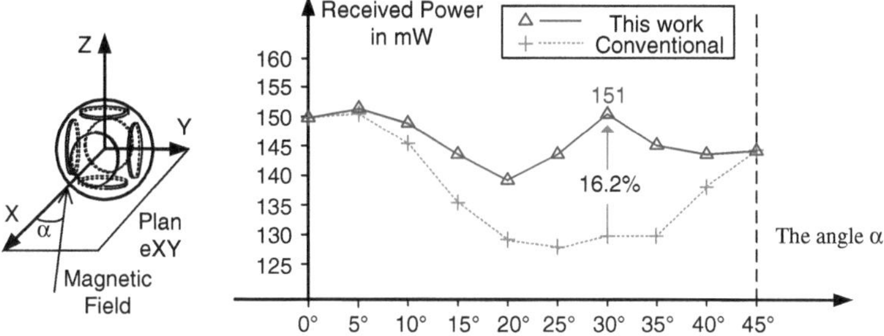

Fig. 6.20 Experimental results

capsule is near the transmitting antenna. In the worst case, the transfer efficiency is around 1.8 % when the capsule is in the center of the body.

6.4 Summing Up

In this chapter, we have proposed two design cases for the batteryless capsule endoscopy. By presenting the two cases, antenna and circuit techniques introduced in the previous chapters are used in practice.

In the first design case, to allow patients walking freely, a two-hop wireless power transfer system for the capsule endoscopy inspection has been proposed. The power is transmitted from the transmitters under the floor, resonantly relayed by the power repeater in the patient's jacket, and finally delivered to the capsule in the patient's body. To enhance the power conversion efficiency, a switch-mode rectifier with current ZCP prediction has been introduced. Its rectification efficiency reaches 93.6 %, which is 13.4 % higher than the best previous designs [17]. A power combination circuit named the Skipping Booster has also been demonstrated, which enhances the efficiency by 18 %. In the best case, 3.0 % of the total DC input power can be transferred and converted to available DC power for the capsule over a distance of 1 m, which is the furthest distance reported in batteryless endoscope applications. Considering the transfer distance is 5–30 times extended and the power efficiency is in inverse cubed law of the distance, the overall power efficiency of 3.0 % is a remarkable result.

In the second design case, a wireless power transfer system that transfers power from a specially designed jacket to the endoscopic capsule in human body is proposed. The jacket uses energy from a group of batteries and transfers the energy to the capsule. Accordingly, this system allows patient going home to operate the endoscopy inspection, which could make patient much more convenient and comfortable. To realize this system, several techniques are proposed to increase the power efficiency. First, a group of segmented antennas are presented to use on the jacket, so the jacket could select the antenna with the highest power transfer efficiency to work. Second, an omnidirectional wireless power receiving circuit is adopted in the capsule, so the capsule can make full use of energy received from every direction. At last, experiments have been conducted. In the best case, the overall transfer efficiency is 5.4 %.

References

1. Xie, X., Li, G. L., Chen, X. K., et al. (2006). A low-power digital IC design inside the wireless endoscopic capsule. *IEEE Journal of Solid-State Circuit, 41*, 2390–2400.
2. Chen, X. K., Zhang, X. Y., Zhang, L. W., et al. (2009). A wireless capsule endoscope system with low-power controlling and processing ASIC. *IEEE Transactions on Biomedical Circuits and Systems, 3*, 11–22.
3. Carpi, F., Kastelein, N., Talcott, M., et al. (2011). Magnetically controllable gastrointestinal steering of video capsules. *IEEE Transactions on Biomedical Engineering, 58*, 231–234.
4. Carta, R., Tortora, G., Thoné, J., Lenaerts, B., Valdastri, P., Menciassi, A., et al. (2009). Wireless powering for a self-propelled and steerable endoscopic capsule for stomach inspection. *Biosensors and Bioelectronics, 25*(4), 845–851.
5. Lenaerts, B., & Puers, R. (2007). An inductive power link for a wireless endoscopy. *Biosensors and Bioelectronics, 22*(7), 1390–1395.
6. Carta, R., Sfakiotakis, M., Pateromichelakis, N., Thoné, J., Tsakiris, D. P., & Puers, R. (2011). A multi-coil inductive powering system for an endoscopic capsule with vibratory actuation. *Sensors and Actuators, A: Physical, 172*(1), 253–258.
7. Puers, R., Carta, R., & Thoné, J., Wireless power and data transmission strategies for next generation capsule endoscopes. *Journal of Micromechanics and Microengineering, 21*, 054008. doi: 10.1088/0960-1317/21/5/054008.
8. Carta, R., Thoné, J., & Puers, R. (2010). A wireless power supply system for robotic capsular endoscopes. *Sensors and Actuators, A: Physical, 162*(2), 177–183.
9. Sun, TJ., Xie, X., & Li, GL., et al. (2011, November). An omnidirectional wireless power receiving IC with 93.6% efficiency CMOS rectifier and skipping booster for implantable bio-microsystems. *A-SSCC* (pp. 185–188).
10. Feng, L., Mao, Y., Cheng, Y. H. (2011, November). An efficiency and stable power management circuit with high output energy for wireless power capsule endoscopy. *A-SSCC*, 229–232.
11. Jourand, P., Carta, R., Puers, R. (2011). Dedicated class-E driver for large area wireless medical inspection capsules. In *Proceedings of Eurosensors XXV Conference, Procedia Engineering* (vol. 25, pp. 1004–1007).
12. Ryu, M., Kim, J. D., Chin, H. U., et al. (2007). Three-dimensional power receiver for in vivo robotic capsules. *Medical and Biological and Engineering and Computing, 45*, 997–1002.
13. Sun, T. S., Xie, X., Li, G. L., et al. (2010, August). An asymmetric resonant coupling wireless power transmission link for micro-ball endoscopy. *EMBC* (pp. 6531–6534).
14. Fang, X., Liu, H., Li, G. Y., et al. (2011, August). Wireless power transfer system for capsule endoscopy based on strongly coupled magnetic resonance theory. *ICMA* (pp. 232–236).
15. Shiba, K., Nagato, T., Tsuji, T., et al. (2008). Energy transmission transformer for a wireless capsule endoscope: analysis of specific absorption rate and current density in biological tissue. *IEEE Transactions on Biomedical Engineering, 55*(7), 1864–1871.
16. Shiba, K., Morimasa, A., & Hirano, H. (2010). Design and development of low-loss transformer for powering small implantable medical devices. *IEEE Transactions on Biomedical Circuits and Systems, 4*(2), 77–85.

17. Lee, S. B., Lee, H. M., et al. (2010, February). An inductively powered scalable 32-channel wireless neural recording system-on-a-chip for neuroscience applications. *ISSCC*, 120–121.
18. O'Driscoll, S., Poon, A. S. Y., Meng, T. H. (2009, February). A mm-sized implantable power receiver with adaptive link compensation. *ISSCC* (pp. 294–295).
19. Harrison, R. R., Watkins, P. T., Kier, R. J., et al. (2007). A low-power integrated circuit for a wireless 100-electrode neural recording system. *IEEE Journal of Solid-State Circuits, 42*, 123–133.
20. Yoo, J., Yan, L., Lee, S., et al. (2009, February). A 5.2 mW self-configured wearable body sensor network controller and a 12 uW 54.9% efficiency wirelessly powered sensor for continuous health monitoring system. *ISSCC* (pp. 290–291).
21. Lin, C. W., Chiu, H. W., Lin, M. L., et al. (2010). Pain control on demand based on pulsed radio-frequency stimulation of the dorsal root ganglion using a batteryless implantable CMOS soc. *ISSCC* (pp. 234–235).
22. Chow, E. Y., Chakraborty, S., Chappell, W. J., et al. (2010, February). Mixed-signal integrated circuits for self-contained sub-cubic millimeter biomedical implants. *ISSCC* (pp. 236–237).
23. McCormick, D., Hu, A. P., Nielsen, P. D., Malpas, S., et al. (2007, August). Powering implantable telemetry devices from localized magnetic fields. *EMBC* (pp. 2331–2335).
24. Finkenzeller, K. (2003). *RFID handbook: fundamentals and applications in contactless smart cards and identification* (2nd ed.). New York: Wiley.
25. Kurs, A., Karalis, A., Moffatt, R., et al. (2007). Wireless power transfer via strongly coupled magnetic resonances. *Science, 317*, 83–86.
26. O'Handely, R. C., Huang, J. K., Bono, D. C., et al. (2008). Improved wireless transcutaneous power transmission for in vivo applications. *IEEE Sensor Journal, 8*, 57–62.
27. Casanoava, J. J., Low, Z. N., & Lin, J. (2009). Design and optimization of a class-E amplifier for a loosely coupled planar wireless power system. *IEEE Transactions on Circuit and System II, 56*(11), 830–834.
28. Low, Z. H., Chinga, R. A., Tseng, R., et al. (2009). Design and test of a high-power high-efficiency loosely coupled planar wireless power transfer system. *IEEE Transactions on Industrial Electronics, 56*, 1801–1812.
29. Casanova, J. J., Low, Z. N., & Lin, J. (2009). A loosely coupled planar wireless power system for multiple receivers. *IEEE Transactions on Industrial Electronics, 56*(8), 3060–3068.
30. Cannon, B. L., Hoburg, J. F., Stancil, D. D., et al. (2009). Magnetic resonant coupling as a potential means for wireless power transfer to multiple small receivers. *IEEE Transactions on Power Electronics, 24*(7), 1819–1825.
31. Sun, T. J., Xie, X., Li, G. L., et al. (2010, December). A wireless energy link for endoscope with end-fire helix transmitter and load-adaptive power converter. *APCCAS* (pp. 32–35).
32. Yuan, Q. W., Chen, Q., Li, L., et al. (2010). Numerical analysis on transmission efficiency of evanescent resonant coupling wireless power transfer system. *IEEE Transactions on Antennas and Propagation, 58*(5), 1751–1758.
33. Chen, C. J., Chu, T. H., Lin, C. L., et al. (2010). A study of loosely coupled coils for wireless power transfer. *IEEE Transactions on Circuit and System II, 57*(7), 536–540.
34. Yoon, I. J., & Ling, H. (2010). Realizing efficient wireless power transfer using small folded cylindrical helix dipoles. *IEEE Antenna and Wireless Propagation Letters, 9*, 846–849.
35. Zhang, F., Hackworth, S. A., Fu, W., et al. (2011). Repeater effect of wireless power transfer using strongly coupled magnetic resonances. *IEEE Transactions on Magnetics, 47*(5), 1478–1481.
36. Fotopoulou, K., & Flynn, B. W. (2011). Wireless power transfer in loosely coupled links: coil misalignment model. *IEEE Transactions on Magnetic, 47*(2), 416–430.
37. Imura, T., Hori, Y. (2011). Maximizing air gap and efficiency of magnetic resonant coupling for wireless power transfer using equivalent circuit and Neumann formula. *IEEE Transactions on Industrial Electronics*, *58*, 4746–4752.

38. Kumar, A., Mirabbasi, S., & Mu, C. (2011). Design and optimization of resonance-based efficient wireless power delivery systems for biomedical implants. *IEEE Transactions on Biomedical Circuits and Systems, 5*(1), 48–63.
39. Ko, W. H., Liang, S. P., Fung, C. D. F. (1977). Design of radio-frequency coils for implant instruments. *Medical and Biological Engineering and Computing*, *15*, 634–640.
40. Terman, F. E. (1943). *Radio engineers handbook.* New York: McGraw-Hill.

Chapter 7
Contributions and Future Work

Abstract To conclude, we have given an introduction to the wireless power transfer for biomedical applications in Chap. 1. By developing the wireless power transfer, researchers hope to provide biomedical devices with more power, longer operating time, smaller system size, and safer insurance. Chapter 2 has presented the systematic design for typical transfer systems. The working principles, classifications, systematic design considerations, and safety concerns were introduced. From Chaps. 3, 4, 5, we have illustrated many designs regarding power antennas, power converters, and power management. By using them, two design cases have been proposed in Chap. 6 for the batteryless capsule endoscopy. In the final chapter, we are going to review the work we have introduced. Moreover, we would look into future and summarize major challenges in the future.

7.1 Main Contributions

The traditional approach to supplying power to biomedical microsystems is to use implantable batteries. However, today's biomedical microsystems have been able to offering much more functions than old times. As a result, the energy budget of implantable batteries severely limits the system characteristics in terms of functions, operating time, resolution, and so on.

Through this book, we have performed a series of researches to study the design of the wireless power transfer for biomedical microsystems. The main contributions can be summarized from three aspects. (1) A systematic level design method has been presented. The most appropriate solution can be found by using this method. (2) We summarized types of power antennas and power circuits. Some of them are proposed by our team, while others are designed by researchers around the world. By summarizing these techniques together, we showed a complete view of the technology of the wireless power transfer. (3) We proposed two wireless

T. Sun et al., *Wireless Power Transfer for Medical Microsystems*,
DOI: 10.1007/978-1-4614-7702-0_7,

power transfer systems for the capsule endoscopy, which is one of the most complex and difficult one in the biomedical applications since the capsule randomly passively moves in digestive track.

7.1.1 Systematic Concerns and Solutions

The first thing to design a wireless power transfer system is to make sure all systematic considerations are clear. The process of thinking introduced in this book can be briefly summarized as Fig. 7.1 shows.

Figure 7.1 describes the system level design of a wireless power transfer system. Since the power transfer efficiency is always important, the key question is that what is the most important concern in system besides the efficiency? Typically, the answers might be full-directionality, ultra small size, moving receiver, long transfer distance, or high transfer power level.

According to each concern, we have introduced state-of-the-art solutions as Fig. 7.1 shows:

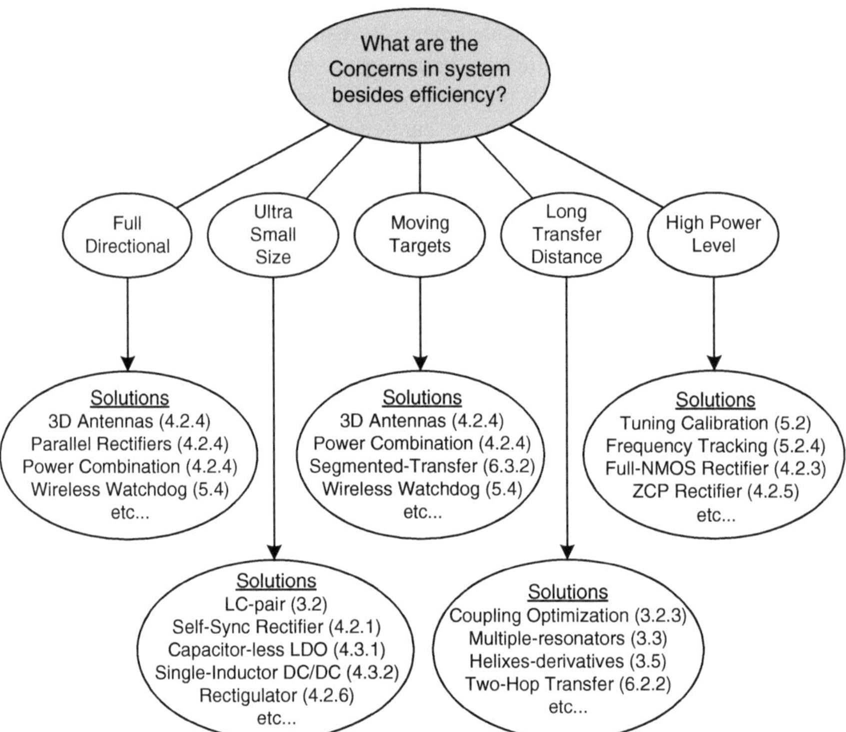

Fig. 7.1 Systematic concerns and solutions

(a) The feature of the full-directionality is very necessary in those microsystems that have uncertain relative angle between the power transmitter and the receiver. The introduced solutions in this book include the 3D antenna (4.2.4), the parallel rectifiers (4.2.4), the power combination (4.2.4), the wireless watchdog for reliability (5.4), and so on. The 3D antennas and parallel rectifier reduce the deviation of power with regard to different relative angles. Since the parallel rectifiers cannot fully utilize the received energy from all directions, the power combination has been presented to improve the efficiency. To ensure the system stability during the transfer, the wireless power watchdog can be adopted. It indicates the received power level, triggers the fail-safe control, and resets the system in the worst case.

(b) For ultra small size systems, typically under several millimeters, high operating frequency can be employed so the antenna size significantly can be reduced. Accordingly, the LC-pair (3.2) can be used to minimize the antenna size. Since the regular rectifiers cannot work at very high frequency, like over GHz, the self-sync rectifier [1] (4.2.1) has been introduced to address this problem. To reduce the size of passive components, the capacitor LDO [2–4] (4.3.1), the single-inductor DC–DC (4.3.2), and the rectigulator (4.2.6) have been presented.

(c) For moving transfer targets, the technical solutions of the 3D antenna (4.2.4), the power combination (4.2.4), the segmented transfer (6.3.2), and the wireless power watchdog (5.4) may be used. The transfer structure of the Multiple-resonators helps decreasing the distance between antennas and increasing the transfer efficiency. The 3D antenna and the power combination ensure system are full-directional. The wireless power watchdog can be used to enhance the system reliability.

(d) For systems requiring long transfer distance, some dedicate antenna techniques including the coupling optimization (3.2.3), the Multiple-resonators (3.3), the helix-derivatives (3.5), and the two-hop transfer (6.2.2) have been introduced. The coupling optimization for LC-pair can be adopted to find the optimized primary coil diameter and maximize the coupling factors between the primary and secondary coils. To extend transfer distance, the Multiple-resonators can be also used, which inserts more resonators in the transmission path. The transfer structure of the helix-derivatives also increases the transfer distance because the adopted high Q factor helical antennas compensate the low coupling factor. The proposed two-hop transfer combines the strong-coupling and the loose-coupling together to deliver power to small biomedical microsystems over long distance.

(e) For systems requiring high transfer power level, the transfer efficiency is more critical than those low-power systems, since more energy would be wasted and there is more harmful radiation. The tuning calibration including the capacitor calibration (5.2.1)–(5.2.3) and the frequency tracking (5.2.4) have been presented for the automatic optimization of tuning circuits, so the coupling factors can be optimized. Besides the coupling efficiency, the circuit's power conversion efficiency can be enhanced by the presented power converters including the full-NMOS rectifier (4.2.3), and the rectifier with ZCP prediction (4.2.5).

Tradeoffs are sometimes needed. For example, the most techniques for the small size feature would cause relatively short transfer distance, small transfer power, and low transfer efficiency. To make an appropriate balance between these techniques is extremely important.

7.1.2 Antennas Summary

There are several ways to summarize the existing antennas introduced in this book. For example, they can be summarized according to their power transfer structures. Or, they can be summarized by the quality factor, maximal efficiency, transfer distance, or applied situation as illustrated in Chap. 3. Here, we try to show a comprehensive summary with respect to the most important features of these antennas. They are summarized in Fig. 7.2.

Total 7 types of antennas are marked in Fig. 7.2. There is no better or worse antenna but only appropriate or not since each antenna technique has unique feature. The LC-pair (3.2) are the most well-known antenna technique, which has relatively small size and simple structure. However, its transfer efficiency rapidly decreases when the transfer distance increases. When the inductor in the LC-pair is achieved by on-chip coils the effective transfer distance is only typically millimeters. Even so, the efficiency is still not satisfying because of its low quality factor. The Multiple-resonators (3.3) are derived from the LC-pair. By inserting more resonators in the transmission path, energy is delivered to farer place. The Helix-derivatives (3.5) is the simplest structure of strong-coupling. It needs only

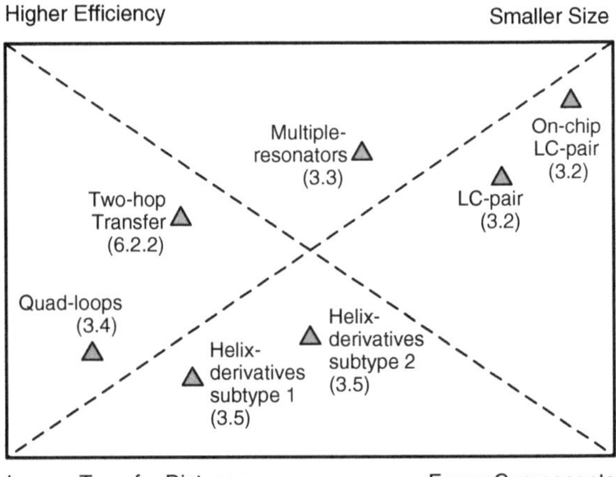

Fig. 7.2 A summary of the introduced power antennas

two helical antennas (subtype 1) or one helical antenna and one magnetic loop coil (subtype 2). By using helical antennas with the quality factor of almost 1,000, the transfer efficiency remains decent over a transfer distance of meters. In the Quad-loops (3.4), two additional loops are used to match the impedance for convenience. The last antenna technique introduced in the book is very special, which is the two-hop transfer (6.2.2). The first hop is realized by strong-coupling and the second hop uses loose-coupling, so it is capable of delivering energy to small receiver over a relatively long distance.

7.1.3 Circuits Summary

This book has also introduced many state-of-the-art circuit designs for the wireless power transfer for biomedical microsystems. Their main functions of these circuits are converting power between DC and AC domains, regulating power in different DC voltages, smartly management the wireless power, and so on. Just like the antenna techniques, there is no better or worse circuits but only appropriate in system or not. Figure 7.3 gives a comprehensive summary of these circuits.

In Fig. 7.3, four characteristics are marked in four corners, and each circuit technique has a position in the figure. By predicting the current zero-cross-point, the rectifier with current ZCP prediction (4.2.5) has the highest power conversion efficiency. The full-NMOS rectifier (4.2.3) also has relatively high efficiency because all power transistors in it are low-resistance NMOS transistors. The comparator-based rectifier [5–7] (4.2.2) is the most frequently adopted rectifier because it has decent efficiency and it's very stable. However, it can be adopted only in those low frequency (under 13.56 MHz) applications. The self-sync rectifier [1] (4.2.1) has very simple circuit structure and it's very suitable for those high-frequency (like over GHz) applications. For those ultra small size

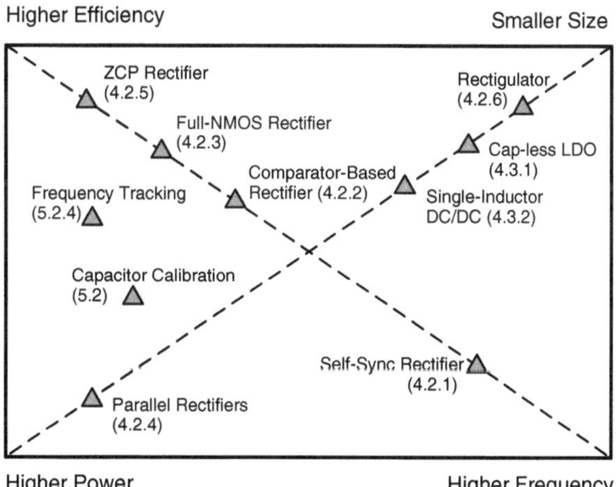

Fig. 7.3 A summary of the introduced power circuits

microsystems, the rectigulator (4.2.6), the capacitor-less LDO (4.3.1), and the multiple-outputs single-inductor DC–DC converter [8–10] (4.3.2) can be used to reduce passive components like off-chip capacitors and inductors. The circuits of the frequency tracking [11, 12] (5.2.4), the capacitor calibration [1, 13] (5.2), and the parallel rectifiers (4.2.4) can be employed in high-power systems. Overall, each circuit has its unique feature. Using Fig. 7.3, designers could find wanted circuit techniques conveniently.

7.2 Future Directions

The wireless power transfer is a rapidly developing technology. According to the current development trend, we believe there are some important future directions. They are the ultra small size receiver for micro implants, the practical middle range transfer, the wireless power transfer with rapid feedback to transmitter, and the wireless power transmitter using energy harvesting. Although these directions emphasize different systematic features, the transfer efficiency is always important.

7.2.1 Ultra Small Size Implants

In the future, biomedical implants will become much smaller, like sub-cubic millimeter [14]. The minimization of medical implants would make patient more conformable, convenient, safer, and make the surgical injury minimized. To design smaller implants, the wireless power receiver in the implants has to be designed much smaller than ever. Both size of the power antennas and the circuits are going to be reduced.

The method to minimize the power antennas might be increasing the operating frequency. By increasing the frequency, the required tuning inductance is decreased. However, whether the operating frequency should be increased is debatable. Some researches [1] claim that a high operating frequency like over GHz should be adopted to optimize transfer efficiency. However, the measured transfer efficiency showed by their research is not satisfying. In their research, 915 MHz is selected as the operating frequency. The diameter of the transmitter is 2 cm, and the diameter of the receiver is 2 mm. As a result, the power transfer efficiency is as low as −33.2 dB (0.05 %) over a distance of 15 mm. We believe the operating frequency should not be primarily determined by the small size receiver, but should be dominated by the large size transmitter, because the transmitter typically has much higher Q factor than the receiver. On one hand, the overall efficiency is determined by the product of the Q factors of the transmitter and the receiver. On the other hand, when the operating frequency alters, the change of the Q factor of the transmitter is usually much more sensitive than the receiver's change. Accordingly, the change of frequency should firstly consider maximizing transmitter's Q factor over receiver's Q factor. Because the diameter of the transmitter is usually much larger than the diameter of the receiver, an optimized frequency will be relatively low. Moreover, the increase of the operating frequency causes more energy losses in human tissue, especially for deep implantations. To sum up, a perfect solution for small size receivers hasn't been found.

The main method to minimize power circuit is to reduce off-chip passive components or integrate them on chip. For example, all rectifiers require at least one decoupling capacitor. Because the decoupling capacitors are usually over 1 nF, they cannot be integrated on chip. We have introduced some circuit techniques like the rectigulator (4.2.6) to address the problem, more circuit technologies or packaging technologies would be proposed in the future.

7.2.2 Middle Range Transmission

One of the most valuable but also the most difficult future directions is the middle range transfer with smaller size antenna and higher power efficiency.

In over 100 years ago, Nikola Tesla tried to deliver wireless power over Atlantic. After that, microwave power or power beam are adopted to successfully transfer power over relatively long distance. However, all these powering techniques are radiative methods. In other words, they have very strong directivity. Consequently, to use these techniques to transfer power to moving targets like a patient in daily life is quite unpractical due to the transmitter has to change its direction to track the receiver. Moreover, these radiative methods are harmful to human body. In 2007, a wireless power transfer system successfully transferred 40 % power over a distance 2 m [15]. It's called the strongly magnetic resonance. The problem is the adopted helical antenna has too large diameter, like in diameter of 0.5 m [15], so it's not practical for medical electronic devices in daily life. Figure 7.4 shows an imagined system in the future.

In the figure, there are one position-fixed transmitter and two moving receivers. It is hoped that the moving receiver 1 has the dimension of around 10 cm, so most of wearable medical device like ECG patch could carry this power receiver. The moving receiver 2 is supposed to have the maximal dimension of 20 cm, so most wearable jacket could carry this receiver. The expected transfer distances for the receivers are 2 and 4 m, so the wireless system could cover at least one room. The expected transfer efficiency is 30 and 60 % respectively. If this imagined system can be realized in the future, the powering of the biomedical devices would be entirely changed.

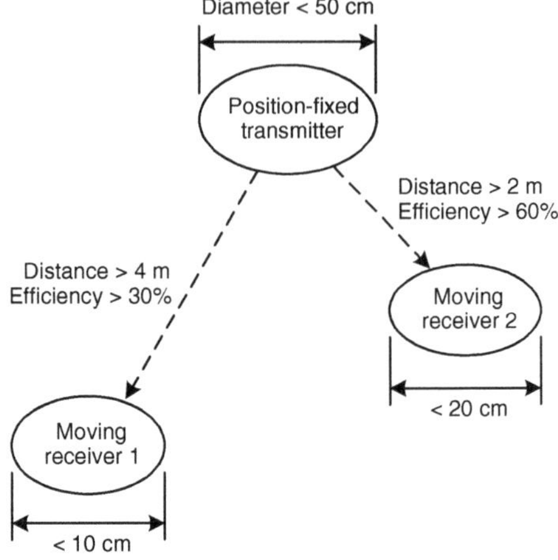

Fig. 7.4 The ultimate challenge for middle range transfer

7.2.3 Transmission with Feedback

One essential difference between wired and wireless power transfers is they have different response from receivers to transmitters. Figure 7.5 shows one regular wired transfer and two wireless power transfers.

Figure 7.5a is a model of a regular wired power transfer. The power demand is feedback from the receiver to the transmitter through the cord. Because of the existence of the cord, the transmitter could monitor the voltage or the current of

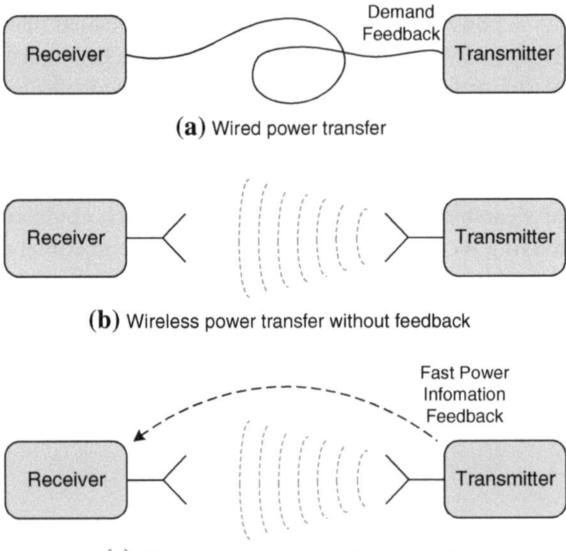

Fig. 7.5 Comparison of wired and wireless power transfers

the receiver, and provide stable voltage or current. By using feedback techniques, the supplied power always meets the power demand. The power transfer is stable, reliable, and highly efficient. Figure 7.5b shows a regular wireless power transfer system, which has no feedback from the receiver to the transmitter. Due to the change of the power demand of the receiver, the transfer distance, the relative angle, and so on, the receiver has to collect energy more than it consumes. Accordingly, the transfer efficiency degrades. Figure 7.5c shows a wireless power transfer with feedback, which is believed one of the future directions. Because of the existence of the feedback, the transmitter could dynamically adjust the power level to satisfy the demand of the receiver. The faster the feedback is, the higher the efficiency is, and more reliable the system is. To establish the feedback, additional circuits and maybe more antennas are required. Moreover, feedback needs extra energy. All these problems are challenging. We believe, in the future, there will be many advanced techniques and even global standards to develop the wireless power transfer with fast feedback from the receiver.

7.2.4 Power Transmitter Using Harvested Energy

Some patients are required to carry implants for lifelong. Although the wireless power transfer has already released patients from changing the implantable battery by surgeries, the patients still have to recharge the power transmitter outside of body or use power cable to connect the transmitter to city power. Since the power consumption of the implanted biomedical microsystem consumes much less current than ever before, we believe one of the future directions is the wearable power transmitter using harvested energy. Figure 7.6 shows the system. By using this system, the user wouldn't even know the existence of the transmitter.

As figure shows, there is a specially designed jacket for patients. The jacket consists of several energy harvesting units, a power transmitter (TX) and a

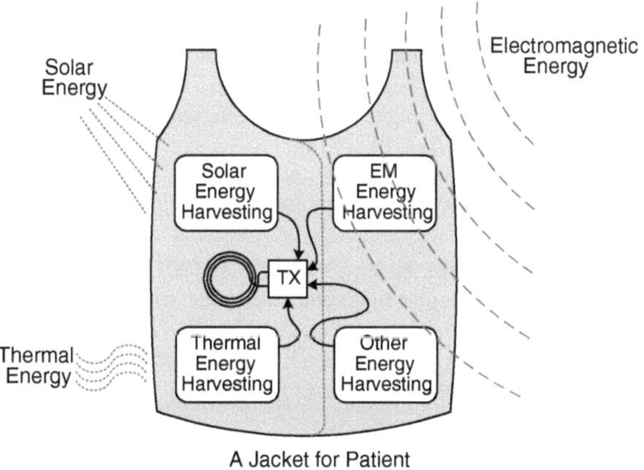

Fig. 7.6 A power transmitting jacket using harvested energy

transmitting coil. The energy harvesting units are responsible to collect kinds of energy, like electromagnetic energy radiated from mobile phone base stations, solar energy, thermal energy, vibration energy, and so on. Because the power of the harvested energy cannot be stable, the harvested energy can be temporary charged in a small-size rechargeable battery in the jacket. When the power of the harvested energy decreases, the battery may discharge to provide enough energy for the implants. Or the harvested energy can be directly sent to the transmitter as real-time power.

Comparing to existing systems, an obvious advantage of this power transmitting jacket is it needs no daily maintenance. No charging or any cable is required for the jacket. As a result, it is very suitable for those patients who require biomedical implants for long term. Patients wouldn't even notice the jacket is special designed.

References

1. O'Driscoll, S., Poon, A., & Meng, T. H. (2009). A mm-sized implantable power receiver with adaptive link compensation, ISSCC (pp. 294–295).
2. Chen, J.-J., Lin, M.-S., Lin, H.-C., et al. (2008). Sub-1 V capacitor-free low-power-consumption LDO with digital controlled loop, APCCAS (pp. 526–529).
3. Kim, Y.-I., & Lee, S.-S. (2012). Fast transient capacitor-less LDO regulator using low-power output voltage detector. *Electronics Letters, 48*(3), 175–177.
4. Chong, S. S., & Chan, P. K. (2012). A 0.9-μA quiescent current output-capacitorless LDO regulator with adaptive power transistors in 65 nm CMOS. *IEEE Transactions on Circuits and Systems I*.
5. Lee, S. B., Lee, H.-M., Kiani, M., et al. (2010). An inductively powered scalable 32-channel wireless neural recording system-on-a-chip for neuroscience applications. *IEEE Transactions on Biomedical Circuits and Systems, 4*(6), 360–371.
6. Sun, Y., Jeong, C., Han, S., et al. (2011). A high speed comparator based active rectifier for wireless power transfer systems, MTT-S (pp. 1–2).
7. Sun, Y., Lee, I., Jeong, C., et al. (2011). A comparator based active rectifier for vibration energy harvesting systems, ICACT (pp. 1404–1408).
8. Chae, C.-S., Le, H.-P., Lee, K.-C., et al. (2009). A single-inductor step-up DC-DC switching converter with bipolar outputs for active matrix OLED mobile display panels. *IEEE Journal of Solid-State Circuits, 44*(2), 509–524.
9. Huang, M.-H., Chen, K.-H., & Wei, W.-H. (2008). Single-inductor dual-output DC-DC converters with high light-load efficiency and minimized cross-regulation for portable devices, VLSI (pp. 132–133).
10. Chang, W.-H., Wang, J.-H., & Tsai, C.-H. (2010). A peak-current controlled single-inductor dual-output DC-DC buck converter with a time-multiplexing scheme, VLSI-DAT (pp. 331–334).
11. Kim, N. Y., Kim, K. Y., Choi, J., et al. (2012). Adaptive frequency with power-level tracking system for efficient magnetic resonance wireless power transfer. *Electronics Letters, 48*(8), 452–454.
12. Fu, W., Zhang, B., & Qiu, D. (2009) Study on frequency-tracking wireless power transfer system by resonant coupling, IPEMC (pp. 2658–2663).

13. Si, P., Hu, A. P., Malpas, S., et al. (2008). A frequency control method for regulating wireless power to implantable devices. *IEEE Transactions on Biomedical Circuits and Systems, 2*(1), 22–29.
14. Chow, E. Y., Chakraborty, S., Chappell, W. J., et al. (2010). Mixed-signal integrated circuits for self-contained sub-cubic millimeter biomedical implants, ISSCC (pp. 236–237).
15. Kurs, A., Karalis, A., Moffatt, R., et al. (2007). Wireless power transfer via strongly coupled magnetic resonances. *Science, 317*(5834), 83–86.

Index

T. Sun et al., *Wireless Power Transfer for Medical Microsystems*,
DOI: 10.1007/978-1-4614-7702-0,

S

T

U

V

W

If you have any concerns about our products,
you can contact us on
ProductSafety@springernature.com

In case Publisher is established outside the EU,
the EU authorized representative is:
Springer Nature Customer Service Center GmbH
Europaplatz 3, 69115 Heidelberg, Germany

Printed by Libri Plureos GmbH
in Hamburg, Germany